Biotechnology and Biochemical Engineering

Prasanna B.D. · Sathyanarayana N. Gummadi
Praveen V. Vadlani
Editors

Biotechnology and Biochemical Engineering

Select Proceedings of ICACE 2015

Springer

Editors
Prasanna B.D.
Department of Chemical Engineering
National Institute of Technology Karnataka
Mangalore, Karnataka
India

Praveen V. Vadlani
Department of Chemical Engineering
Kansas State University
Manhattan, KS
USA

Sathyanarayana N. Gummadi
Bhupat and Jyoti Mehta School
of Biosciences
Indian Institute of Technology Madras
Chennai, Tamil Nadu
India

ISBN 978-981-10-1919-7 ISBN 978-981-10-1920-3 (eBook)
DOI 10.1007/978-981-10-1920-3

Library of Congress Control Number: 2016947027

Printed on acid-free paper

This Springer imprint is published by Springer Nature
The registered company is Springer Science+Business Media Singapore Pte Ltd.

Preface

The Department of Chemical Engineering, National Institute of Technology Karnataka, Surathkal, India was started in the year 1965 with the assistance of UNESCO, MHRD-Government of India, and the Government of Karnataka and is consistently ranked as one of the top chemical engineering departments in India by the National Board of Accreditation. The department has completed 50 years of its glorious service to the cause of Chemical Engineering Education, Research and Industrial collaboration, and celebrated Golden Jubilee Year in 2015. In order to commemorate this important milestone, International Conference on Advances in Chemical Engineering (ICACE)—2015 was organized during December 20–22, 2015.

Chemical engineering is a branch of science that combines physical sciences (physics and chemistry) and life sciences (molecular and synthetic biology) with applied mathematics and economics to synthesize and transform chemicals, materials, and energy from various feedstocks. Essentially, chemical engineers conceptualize novel and innovative processes and implement large-scale industrial operations to produce chemicals and materials suited to every need of the society, including food, chemicals, energy, environment, and information. ICACE-2015 conference provided a platform for deliberating advances in chemical engineering as well as allied fields, through invited lectures delivered by researchers of national and international repute, and paper and poster presentation by participants from academia, research organizations, and industries.

ICACE-2015 had solicited technical and research submissions related to Industrial Biotechnology and Biochemical Engineering. Bioprocessing of renewable resources to useful products is a sustainable approach and has the wherewithal to solve some of the challenging issues facing humanity. We had an overwhelming response to our call for paper submission, and all the submitted papers were screened and selected by the technical committee on the basis of originality, significance, coherence of data and clarity. Altogether 53 papers were received; out of which 37 papers were selected for oral presentation in the conference. The papers presented during the conference were again subjected to peer-review and plagiarism check. Based on the outcome, 26 papers have been selected for publication in this book.

We would like to thank Dr. Keyur Raval for helping us in the peer-review process. Our heartfelt thanks to Dr. Kalaivani, Mrs. Jaya Mary, and Mr. Basavaraj Nainegli for their editorial assistance. We are highly grateful to NITK administration, TEQIP, and technical committee members for their cooperation. The efforts by faculty, staff, students, and alumni of NITK Surathkal in organizing ICACE-2015 are greatly appreciated. We are grateful to all those who have contributed directly and indirectly for the success of this event. We hope that all the readers will find this book volume useful and productive.

Karnataka, India Prasanna B.D.
Chennai, India Sathyanarayana N. Gummadi
Manhattan, USA Praveen V. Vadlani

Contents

About the Editors

Dr. Prasanna B.D. is Assistant Professor in the Department of Chemical Engineering, National Institute of Technology Karnataka. He has produced more than 25 research publications in the field of biotechnology and two patent applications. His research areas include fermentation process development: enzymes, probiotics and nematodes; downstream process technologies; and oenology.

Prof. Sathyanarayana N. Gummadi is Professor in the Department of Biotechnology, Bhupat and Jyoti Mehta School of Biosciences, Indian Institute of Technology Madras. His research primarily focuses on understanding the molecular and physiological basis of survival of microbes under extreme conditions and exploiting their industrial importance, and on understanding the molecular and biochemical basis of phospholipid translocation in cellular membranes.

Dr. Praveen V. Vadlani is the Gary and Betty Lortscher Renewable Energy Associate Professor at the Department of Grain Science and Industry, Kansas State University, USA. He has more than 30 research publications to his credit and is currently working in the advanced chemical engineering field.

Selection of Medium and Optimization of Process Parameters for Melanin Biosynthesis from *Pseudomonas stutzeri* HMGM-7

Harsha Thaira, Shraddha S. Bhosle, Rajmohan Balakrishnan and Keyur Raval

1 Introduction

Pigments are colorful chemical compounds which absorbs light in the visible spectrum. The produced color is due to the absorption of energy by a group of molecules known as chromophore which leads to the excitation of an electron. The non-absorbed energy which is reflected or refracted is captured by the eye to generate neural impulses which are then carried to the brain where they could be decoded as a color (Hari 1994). Pigments are of two types, synthetic pigments and natural pigments. They are widely used in clothes, cosmetics, furniture, foods, medicines, and in other products. Based on their structural characteristics, the natural pigments are classified as Tetrapyrrole derivatives, Isoprenoid derivatives, Benzopyran derivatives, Quinones and Melanins (Hari 1994). Melanins are nitrogenous polymeric compounds with indole ring as their monomeric unit but they are not homopolymers. Generally, they are present as a mixture of macromolecules and are responsible for most of the black, brown and gray colorations of plants, animals, and microorganisms. Melanins are classified into three groups, they are; Eumelanins which are black or brown pigments and are widely distributed in vertebrates and invertebrates. It is the most common type of melanin. Pheomelanins which are yellow to red pigments and are found in mammals and birds. And finally allomelanins that are present in fungi, seeds and spores.

H. Thaira (✉) · S.S. Bhosle · R. Balakrishnan · K. Raval
Department of Chemical Engineering, National Institute of Technology Karnataka, Mangaluru, Karnataka, India
e-mail: harshathaira@gmail.com

Prasanna B.D. et al. (eds.), *Biotechnology and Biochemical Engineering*,
DOI 10.1007/978-981-10-1920-3_1

The very first report of melanin production was reported in *Pseudomonas aeruginosa* producing pyomelanin (Osawa et al. 1963) followed by *Shewanella colwelliana*, *Vibrio cholera*, and *Hyphomonas* strain (Kotob et al. 1995; Ruzafa et al. 1995). The melanin synthesis using homogentisic acid as a precursor was first reported in *Vibrio cholerae*, *Hyphomonas* species and *Shewanella colwelliana* (Kotob et al. 1995). The synthesis of melanin and its characterization such as solubility, free radical nature was initially studied in *Proteus mirabilis* (Agodi et al. 1996). A novel marine bacterium *Alteromonas* strain MMB-1, was isolated from the Mediterranean Sea and its melanin synthesis ability was studied using L-tyrosine as a precursor (Solano et al. 1997). Melanin production was studied by UV-resistant mutant of *Bacillus thuringiensis* subsp. *Kurstaki* and its UV-protection ability for insecticidal crystals was tested (Saxena et al. 2002). The thermo tolerant strains of *Bacillus thuringiensis* were also reported for melanin production (Ruan et al. 2004). Another important melanin producing bacterium reported was *Marinomonas mediterranea*, which produces black eumelanin from L-tyrosine (Lucas-Elio et al. 2012). A marine bacterium, *Pseudomonas stutzeri* HMGM-7 producing considerable amount of melanin in sea water medium without the addition of L-tyrosine was also reported (Ganesh Kumar et al. 2013).

The different properties and functions of melanins are being explored for various applications. The high reactivity of melanin due to the presence of =O, –OH, –NH, and –COOH groups is the reason behind growing interest in melanin research. Melanins have a broad spectrum of biological applications. This includes inhibition of Human Immunodeficiency Virus (HIV) replication, antivenin activity, antimicrobial activity and antioxidant activity. Melanins also have physical and chemical applications. This includes their use in nano particle synthesis, cosmetics and in lenses.

Inspite of so many potential applications, melanins are not used due to non-availability of a sustainable and cost effective method of melanin production. The major sources of melanin are cephalopods, plants and microorganisms. Melanin obtained from microbes has great advantages over melanin from animals and plants. Microorganisms do not cause the problems of seasonal variations and are fast in growth. They also modify themselves according to the medium and conditions provided. The bacterial sources have been used as a main source of melanin with immense applications in the field of agriculture, cosmetics and pharmaceutical industries (Riley 1997), hence its optimization is important for large scale production. The high production cost and high commercial value of melanin has given rise to the need of a demanding research for cheaper production methods. The present study aims to increase the productivity of melanin from an epiphytic bacterium known as *Pseudomonas stutzeri* HMGM-7 (MTCC 11712). This bacterium was isolated from the branches of sea weed *Hypnea musciformis*, which released a black extracellular pigment into the medium. The physical parameters and nutritional requirements which plays an important role in the production of melanin were optimized in the present study.

2 Materials and Methods

2.1 *Microorganisms and Culture Conditions*

Nutrient agar slants and plates were prepared for maintenance of the organism (*Pseudomonas stutzeri* HMGM-7) obtained from MTCC Chandigarh. Periodical subcultures were done for maintenance of the viability of the strain. Media volume used throughout the experiments was 50 ml in 250 ml Erlenmeyer flasks which was maintained at 37 °C at 150 rpm. Nutrient broth prepared in distilled water was used for shake flask studies.

2.2 *Growth Studies*

2 % inoculum volume was added to each flask and incubated at 37 °C and observed at two different rpm: 150 and 250. During the incubation period of 72 h, the OD of the samples was measured at 660 nm after every 4 h using appropriate blank. Biomass dry weight for a volume of 28 ml of the culture was accounted for by centrifuging (8000 rpm, 8 min, and 40 °C) and drying the pellet for 8 h at 60 °C in a hot air oven. The supernatant obtained after centrifugation is filter sterilized with 0.45 μm syringe filters and their absorbance is measured at 400 nm to quantify the melanin.

2.3 *Optimization of Nutritional Parameters for Melanin Production*

All the experiments conducted used a constant media volume of 50 ml in 250 ml Erlenmeyer flasks. NB was used as the basal medium which was incubated at 37 °C and 150 rpm unless otherwise stated. The effect of different parameters on the production of melanin such as pH, temperature, carbon sources, nitrogen sources, trace elements were evaluated by keeping Nutrient broth as the basal medium. The factorial design of experiments known as ‘one factor at a time’ method was applied to improve the reproducibility of the experimental results and to optimize the entire biosynthesis process. The experiments were conducted by varying one factor at a time and keeping the remaining factors constant.

2.4 *Effect of Inoculum Age*

The effect of age of the inoculum on the melanin production was studied using Nutrient Broth medium by using 6, 12, 18 and 32 h old cultures maintained at 37 °C and 150 rpm.

2.5 Effect of Inoculum Size

0.5, 5, 10, 15 and 20 % inoculum volume were evaluated for melanin production using Nutrient Broth medium at 37 °C and 150 rpm.

2.6 Effect of Shaking Frequency

The shaking frequency was optimized by incubating the Erlenmeyer flasks in incubator shaker at 100, 150, 200 and 250 rpm at 37 °C with a shaking diameter of 25 mm.

2.7 Effect of pH

The optimum pH for the production of melanin was determined by setting initial pH of medium to 4, 5, 6, 7, 8 and 9 by using 0.1 N HCl and 0.1 N NaOH.

2.8 Effect of Temperature

The optimization of temperature for melanin production was carried out by incubating flasks at 30, 35, 37, 40 and 45 °C in incubator shaker.

2.9 Effect of Different Growth Media

Four different growth media, Nutrient Broth (NB), Luria Bertini (LB) Broth, Bushnell-Haas Broth (BHB) and Trypticase Soy Broth (TSB) were studied for melanin production by culturing the microorganism in each of the medium at 37 °C and 150 rpm.

2.10 Effect of Carbon Sources

The effect of various carbon sources was studied by adding each carbon source in the medium at the concentration (2.5 g/L). The carbon sources evaluated were glucose, sucrose, lactose, fructose, starch, xylose, maltose, glycerol and dextrose.

2.11 *Effect of Organic Nitrogen Sources*

To evaluate the various nitrogen sources for maximum melanin, the production medium was supplemented with each organic nitrogen source at the concentration (1.5 g/L). The organic nitrogen sources tested were peptone, beef extract, yeast extract, and tryptone.

2.12 *Effect of L-Tyrosine*

To study the effect of L-tyrosine on melanin yield, 1.6 g/L of L-tyrosine was added to TSB and NB and were incubated in an orbital shaker at 150 rpm maintained at 37 °C.

2.13 *Extraction and Purification of Melanin*

The extraction of melanin was done in accordance with the procedure described for the purification of melanin from the culture of *Aspergillus bridgeri* (Kumar et al. 2011) with some minor modifications. In short, the medium was centrifuged at 5000 g for 10 min to remove the biomass. The supernatant collected was then treated with 1 M NaOH and then autoclaved at 120 °C for 15–20 min. After autoclaving, the solution was cooled and centrifuged at 5000 g for 10 min to collect the alkylated supernatant which was then acidified to pH 2 by using 1 N HCl, in order to precipitate the melanin. The precipitated melanin was collected by centrifuging at 12,000 g for 20 min and washed with distilled water and evaporated to dryness at room temperature and was stored for further use.

2.14 *Characterization Studies*

Purified melanin was dissolved in 0.1 N NaOH for UV-visible spectrophotometric analysis. The solution was scanned from 200 to 900 nm. The absorbance was measured by using a double beam UV-visible spectrophotometer (Hitachi, Labomed Inc). The absorption spectrum of the melanin pigment from the *Pseudomonas stutzeri* HMGM-7 strain was compared with that of standard melanin. For FT-IR analysis, the pigment and standard melanin were scanned between the wavenumber range of 4000–400 cm^{-1} by using KBr discs with an FT-IR spectrophotometer (IR Prestige, Shimadzu).

3 Results and Discussions

3.1 *Effect of Inoculum Age*

After evaluating different inoculum age (6, 12, 18, and 32 h) for their melanin production, the 12 h old inoculum gave the highest melanin concentration (197 mg/L) at the 48th h. When 6 h old culture was used, the highest melanin concentration (195.2 mg/L) was attained in the 40th h itself. The maximum biomass yield was obtained for the 32 h old culture (1.434 g/L 12th h) whereas the 6 h old culture managed to attain its maximum biomass in the 22nd h (1.69 g/L). 1.344 and 1.410 g/L were the highest biomass concentrations for cultures that were 12 and 18 h old respectively. The highest melanin obtained in the control medium was 177 mg/L and biomass attained was 1.107 g/L. Increase in inoculum age thus results in increase in biomass whereas reduction inoculum age resulted in increase in melanin production.

3.2 *Effect of Inoculum Size*

Different inoculum volumes (0.5, 5, 10, 15 and 20 %) were investigated to observe their effect on melanin production. The highest melanin production of 270.9 mg/L (32nd h) was achieved when 10 % inoculum was added to 50 ml of Nutrient Broth medium. Change in inoculum size did not alter the biomass yields, 1.425, 1.4, 1.46, 1.418 and 1.385 g/L being the maximum biomass concentrations obtained at 0.5, 5, 10, 15 and 20 % inoculum sizes respectively. As a result of which, 10 % inoculum volume was selected to be the optimum inoculum volume to be used for further investigations. The melanin production was observed after 10th h. The highest yield of melanin was achieved (270.9 mg/L) at the 32nd h. Thus there was a 53 % increase in melanin productivity.

3.3 *Effect of Shaking Frequency on Biomass Production*

Nutrient broth medium was used as the growth medium which was maintained at 37 °C and the shaking frequency was varied to study its effect on the melanin production. The maximum biomass production obtained for *Pseudomonas stutzeri* HMGM-7 was 1.107 g/L at the 8th and 12th h when the organism was allowed to grow for a period of 72 h at 150 rpm. In the medium prepared in sea water without adding L-tyrosine, Ganesh Kumar et al. (2013) obtained maximum biomass production of 2.5 g/L. There was a substantial increase in the biomass at the increased shaking frequency of 250 rpm as compared to 150 rpm.

3.4 Effect of Shaking Frequency on Melanin Production

There was a steady increase in the melanin production till the 48th h when the culture flasks were maintained at 37 °C and 150 rpm, where maximum production of 0.177 g/L was obtained followed by a decline in its productivity by the end of the incubation period. The onset of melanin production was significant only after the 8th h. Ganesh Kumar et al. (2013) obtained the maximum melanin production of 6.7 g/L at the 60th h in the sea-water medium without L-tyrosine supplementation. The melanin yield obtained in this study is comparatively lesser since Nutrient broth medium prepared in distilled water was used instead of sea-water medium which is known to be conducive for marine species like *Pseudomonas*. Melanin production at 37 °C and 250 rpm increased till the 32nd h, where maximum production of 0.164 g/L was obtained followed by a decline in its productivity by the end of the incubation period. When L-tyrosine was used as a sole carbon and nitrogen source into the melanin production media containing KH_2PO_4, NaCl and $MgSO_4 \cdot 7H_2O$ made in Distilled water by Kurian et al. (2014), *Pseudomonas stutzeri* Strain BTCZ10 produced 47.47 ± 0.2 μg/mL of melanin. Thus, in present study it was found that increase in shaking frequency from 150 to 250 rpm caused a decrease in melanin productivity.

3.5 Effect of Growth Media

Nutrient Broth (NB), Luria Bertini (LB) broth, Bushnell-Haas broth (BHB) and Trypticase Soy broth (TSB) were the four different growth media that were utilized in this study to screen for the medium that produced more amount of melanin. The highest melanin yield, 167.38 mg/L was obtained at the 32nd h in NB, followed by TSB and LB, whereas BHB had very low melanin production (Fig. 1). None of the additional nutrients could affect a significant rise in melanin production when compared to NB alone.

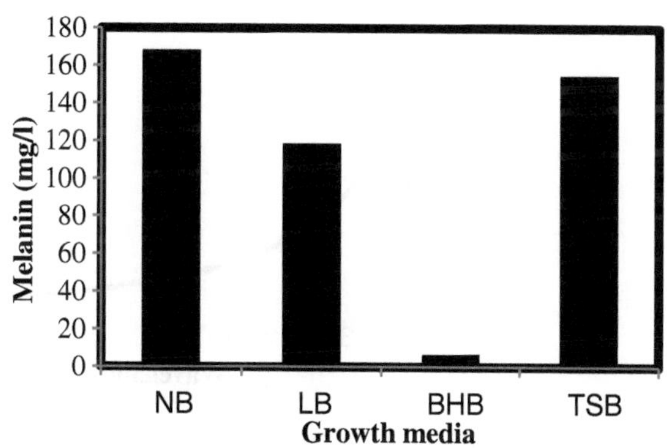

Fig. 1 Melanin productions for different growth media

3.6 Melanin Yield Before and After Optimization

The highest yield of melanin concentration, 0.27 g/L, was obtained in nutrient broth at 32 h. The yield was 1.53 times higher than the melanin obtained before optimization, 0.177 g/L at 48 h (Fig. 2). The increase in the productivity of melanin after selection of suitable medium and optimization of process parameters was 128.73 %. The melanin yield obtained can be further enhanced by statistical optimization and evaluating the effect of different combinations of nutrients like carbon and nitrogen sources and trace elements and further scale up of the process can be done.

3.7 UV-Visible Absorption Spectra and FTIR Analysis

The spectral property of the pigment was analyzed to confirm the nature of the pigment. Its UV spectrum was found to be similar to that of synthetic melanin which exhibited absorption peak of maxima between 200 and 300 nm (Fig. 3).

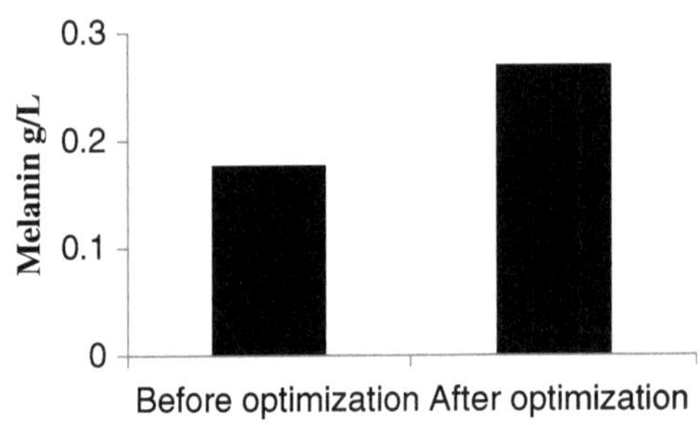

Fig. 2 Melanin yield before and after optimization

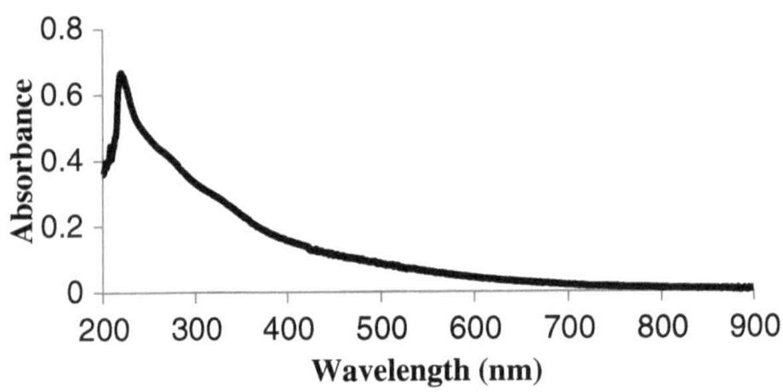

Fig. 3 The UV absorption spectrum of melanin produced by *Pseudomonas stutzeri* HMGM-7

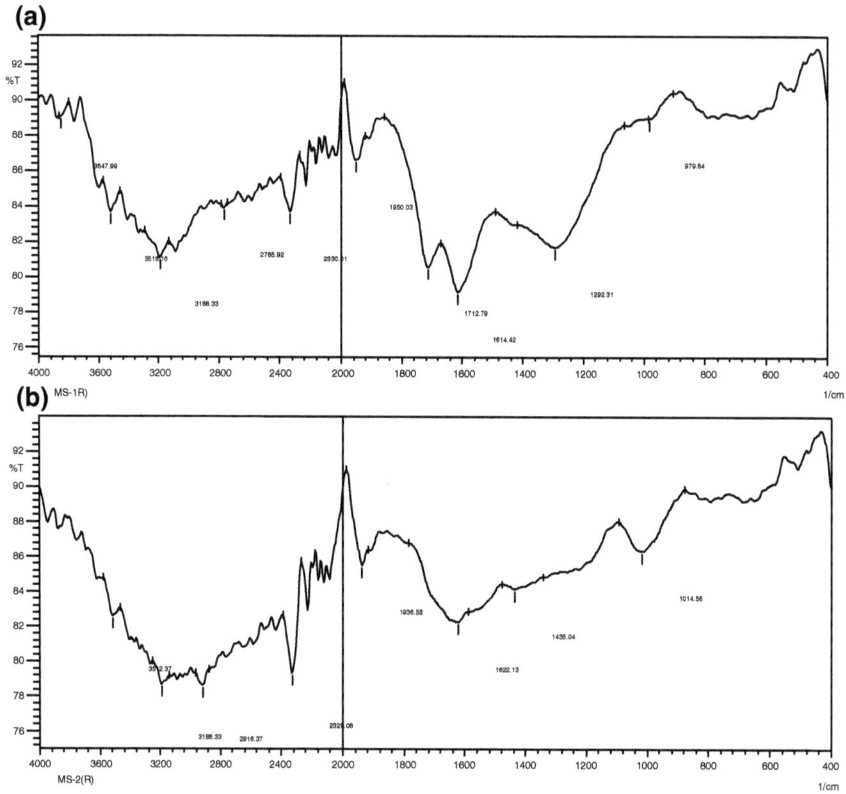

Fig. 4 **a** FTIR spectra of synthetic melanin, **b** FTIR spectra of melanin obtained from *Pseudomonas stutzeri* HMGM-7

It showed a high degree of similarity, when the main absorption peaks in the FT-IR spectra of the synthetic melanin (Sigma Aldrich) and melanin obtained from *Pseudomonas stutzeri* HMGM-7 were compared (Fig. 4).

4 Conclusion

Many bacterial sources have been used widely as a major source of melanin in recent years and hence its optimization is important for large scale production. *Pseudomonas stutzeri* HMGM-7 used in this investigation has a competence to produce melanin under various process conditions and in different growth medium and can prove to be of commercial use for the large scale industrial production. Physical parameters and nutritional requirements often determine the melanin productivity that can be obtained from bacterial sources, and hence these parameters were evaluated in the current study. *Pseudomonas stutzeri* HMGM-7 was able to produce its highest melanin yield within shorter incubation period (32 h) for most of the studies that were conducted by varying different nutritional and process parameters. The optimum inoculum age and size that produced higher melanin yield was found to be 6 h and 10 % respectively. Nutrient broth along with three different media (TSB, BHB, LB) were evaluated for its melanin production, from which Nutrient broth proved to be the best, 0.27 g/L being the highest melanin yield produced across all the experiments conducted. There was no significant increase in the melanin production when the media was supplemented with additional nutrients.

References

Agodi, A., Stefani, S., Corsaro, C., Campanile, F., Gribaldo, S., Sichel, G.: Study references: of a melanic pigment of *Proteus mirabilis*. Res. Microbiol. **147**(3), 167–174 (1996)

Ganesh Kumar, C., Mongolla, P., Pombala, S., Kamle, A., Joseph, J.: Physicochemical characterization and antioxidant activity of melanin from a novel strain of *Aspergillus bridgeri* ICTF-201. Lett. Appl. Microbiol. **53**(3), 350–358 (2011)

Ganesh Kumar, C., Sahu, N., Narender Reddy, G., Prasad, R.B., Nagesh, N., Kamal, A.: Production of melanin pigment from *Pseudomonas stutzeri* isolated from red seaweed *Hypnea musciformis*. Lett. Appl. Microbiol. **57**(4), 295–302 (2013)

Hari, R.K., Patel, T.R., Martin, A.M.: An overview of pigment production in biological systems: functions, biosynthesis, and applications in food industry. Food Rev. Int. **10**(1), 49–70 (1994)

Kotob, S.I., Coon, S.L., Quintero, E.J., Weiner, R.M.: Homogentisic acid is the primary precursor of melanin synthesis in *Vibrio cholerae*, a Hyphomonas strain, and *Shewanella colwelliana*. Appl. Environ. Microbiol. **61**(4), 1620–1622 (1995)

Kurian, N.K., Nair, H.P., Bhat, S.G.: Melanin producing *Pseudomonas stutzeri* BTCZ10 from marine sediment at 96 m depth (Sagar Sampada cruise #305). Int. J. Curr. Biotechnol. **2**(5), 6–11 (2014)

Lucas-Elio, P., Goodwin, L., Woyke, T., Pitluck, S., Nolan, M., Kyrpides, N.C., Sanchez-Amat, A.: Complete genome sequence of the melanogenic marine bacterium *Marinomonas mediterranea* type strain (MMB-1(T)). Stand Genomic Sci. **6**(1), 63–73 (2012)

Osawa, S., Yabuuchi, E., Narano, Y., Nakata, M., Kosono, Y., Takashina, K., Tanabe, T.: Pigment production by *Pseudomonas aeruginosa* on glutamic acid medium and gel filtration of the culture fluid filtrate. Jpn. J. Microbiol. **7**, 87–95 (1963)

Riley, P.: Melanin. Int. J. Biochem. Cell Biol. **29** (1997)

Ruan, L., Yu, Z., Fang, B., He, W., Wang, Y., Shen, P.: Melanin pigment formation and increased UV resistance in *Bacillus thuringiensis* following high temperature induction. Syst. Appl. Microbiol. **27**(3), 286–289 (2004)

Ruzafa, C., Sanchez-Amat, A., Solano, F.: Characterization of the melanogenic system in *Vibrio cholerae*, ATCC 14035. Pigment Cell Res. **8**(3), 147–152 (1995)

Saxena, D., Ben-Dov, E., Manasherob, R., Barak, Z., Boussiba, S., Zaritsky, A.: A UV tolerant mutant of *Bacillus thuringiensis* subsp. kurstaki producing melanin. Curr. Microbiol. **44**(1), 25–30 (2002)

Solano, F., Garcia, E., Perez, D., Sanchez-Amat, A.: Isolation and characterization of strain MMB-1 (CECT 4803), a novel melanogenic marine bacterium. Appl. Environ. Microbiol. **63** (9), 3499–3506 (1997)

Unstructured Kinetic Modeling of Glutathione Production by *Saccharomyces cerevisiae* NCIM 3345

Abhinandan Dhavale, Atul Vhanmarathi, Shrinivas Deshmukh and Seema Dabeer

1 Introduction

Glutathione (GSH) is a biologically active tri-peptide consisting of L-glutamate, L-cysteine, and glycine. It is an abundant and ubiquitous low-molecular-mass thiol in living tissues. GSH is an important component in the cellular mechanisms that protect against UV (Sollod et al. 1992), heavy metals (Perego et al. 1997), and many exogenous organic substances (Goto et al. 1995). GSH also plays a very critical role in sacrificial defensive mechanism against oxidative damage in organisms (Berhane et al. 1994). It is now widely used as a medicine, in health foods to prevent hepatotoxicity induced by acetaminophen, vinyl ethers or bromobenzene in animals and in the cosmetic industry. Glutathione majorly functions as an antioxidant, an immunity booster, and as a detoxifier (Pastore et al. 2003).

Recently several studies have described GSH producing yeast strains, which are commonly used for commercial production such as *Saccharomyces cerevisiae* and *Candida utilis* (Sakato and Tanaka 1992; Wei et al. 2003a, b). Many studies have tried to improve the GSH production by supplementing certain materials, such as glucose, minerals, ATP, and phosphorus to the culture medium (Li et al. 1998). The studies have also been reported for investigating the different concentration of amino acids and related compounds such as cysteine on glutathione production (Alfafara et al. 1992). Finally, optimization of fermentation process parameters and conditions were achieved so as to enhance the GSH production using *S. cerevisiae* (Cha et al. 2004).

A. Dhavale (✉) · A. Vhanmarathi · S. Deshmukh · S. Dabeer
K.I.T.'s College of Engineering, Kolhapur 416006, Maharashtra, India
e-mail: abhibiocpd@gmail.com

Prasanna B.D. et al. (eds.), *Biotechnology and Biochemical Engineering*,
DOI 10.1007/978-981-10-1920-3_2

As mentioned above, many kinds of literature provide information about the optimization of the fermentation process for GSH production using *S. cerevisiae*, but very few references in the literature provide information about kinetic modeling of the production processes. It is essential to have an in-depth understanding of the fermentation process so as to establish better controls at commercial scale production to achieve a consistent and quality product. The already known and well-established kinetics models help develop these strategies (Ghaly et al. 2005). In the present study GSH, production was carried out under the optimal condition and various renowned unstructured kinetic models were used to understand the kinetics of GSH production.

2 Materials and Methods

2.1 Microorganism

Saccharomyces cerevisiae NCIM 3345 was procured from NCL Pune, India. While all the media components (LR grade) were bought from Hi-media Ltd, Mumbai, India.

2.2 Culture Maintenance

Saccharomyces cerevisiae NCIM 3345 stock culture was maintained using nutrient media containing; Peptone 10 (g/L), Beef extract 3 (g/L), NaCl 5 (g/L). This culture was used to inoculate seed medium with working volume 100 mL containing Glucose 1 g, Peptone 0.5 g, Yeast extract 0.3 g and Malt extract 0.3 g. Seed flask i.e. inoculum was incubated at 30 °C for 24 h.

2.3 Fermentation

The GSH formation using *Saccharomyces cerevisiae* NCIM 3345 was achieved in the same 100 mL production medium in 250 mL Erlenmeyer flask containing these components in (g/L): Glucose 30, Yeast extract 30, KH_2PO_4 0.6 and L-cysteine 0.6. The pH of media was set to 5.5 and further subjected to sterilization using autoclave at 121 °C for 30 min. Post sterilization, the flasks were inoculated with 8 % (v/v) of matured seed obtained from seed flask. To achieve desired growth and product formation, production flasks were incubated in an orbital incubator at 30 °C with a speed of 150 rpm for 16 h. Sampling was performed as per the protocol for estimation of biomass growth, glutathione production, and glucose utilization. All the analysis was carried out in triplicate.

2.4 *Biomass Determination*

The concentration of cell biomass was obtained by calculating dry cell weight (DCW) of the cells. Eight mL broth was taken in pre-weighted centrifuge tubes and was subjected to centrifugation at 5000 rpm for 10 min at room temperature. Supernatant was discarded and the pellet was washed using distilled water (twice). The tubes containing pellets were further kept in hot air oven at 100 °C for drying. The weight of each tube was checked after 24 h of drying, ensuring all moisture had evaporated.

2.5 *Glucose Determination*

To calculate the residual substrate concentration i.e. glucose in the production flask, the 3,5-Dinitrosalicylic acid method was used (Miller 1959).

2.6 *Glutathione (GSH) Determination*

After 16 h of fermentation, broth was subjected to centrifugation to achieve biomass pellets, which were further suspended in 0.2 M phosphate buffer (pH 7.2) and disrupted by sonication. The glutathione concentration in the supernatant achieved after centrifugation was measured using published methods (Cohn and Lyle 1966), such as measuring the absorbance maxima (λ_{max}) at 412 nm using a UV spectrophotometer (Thermo-Scientific CHEMITO 215D). A standard curve was generated by preparing samples of different known concentrations of GSH, which was used to determine specimen concentrations.

2.7 *Fermentation Description Using Unstructured Kinetics Models*

2.7.1 Biomass Growth

Various unstructured models had characterized the cell culture growth patterns in fermentation kinetics. In an unstructured model, the total biomass concentration (whole quantity) is considered as a single component in representations (Chandrasekhar et al. 1999). Many mathematical equations and theories are available in literature which can explain the sigmoidal relationship between the specific growth rate of cells and key limiting substrate used for biomass growth. In current study, the most suitable kinetic model tested for describing cell growth was logistic equation.

$$\frac{dX}{dt} = \mu_0\left(1 - \frac{X}{X_{\max}}\right)X \tag{1}$$

where μ_0 are initial specific growth rate (h^{-1}) and $X_{\max}$ maximum attainable biomass concentration (g/L), which on integration, with the initial condition that at t = 0, X = X_0, yields

$$\ln\frac{X}{(X - X_{\max})} = \mu_{\max}t + \ln\frac{X_0}{(X_{\max} - X)} \tag{2}$$

On rearrangement, and explicit function for biomass is obtained as:

$$X_t = \frac{X_0 e^{(\mu_0 t)}}{1 - \frac{X_0}{X_{\max}}(1 - e^{(\mu_0 t)})} \tag{3}$$

2.7.2 Glutathione Production Kinetics

The GSH production kinetics was analyzed according to the Luedeking-Piret equation. This model illustrates that the rate of product formation is mainly dependent on; (1) the desired cell mass for product formation (*X*) and (2) the rate at which biomass is increasing with respect to time (d*X*/d*t*) in a linear fashion.

$$\frac{dP}{dt} = \alpha\frac{dX}{dt} + \beta X \tag{4}$$

In Eq. (4), α and β are the constants determined experimentally, which provides the basis for classification of microbial metabolites into growth associated (β = 0), non-growth associated (α = 0), and mixed (α ≠ 0 and β ≠ 0). Integration of Eq. (4) using Eq. (3) and initial conditions, (X_0, P_0) yields,

$$P_t = P_0 + \alpha A(t) + \beta B(t) \tag{5}$$

In Eq. (5) A (t) and B (t) as follows,

$$A(t) = \left[\frac{X_0 e^{\mu_0 t}}{1 - \left(\frac{X_0}{X_{\max}}\right)(1 - e^{\mu_0 t})} - 1\right] \quad \text{and} \quad B(t) = \frac{X_{\max}}{\mu_0}\ln\left(1 - \frac{X_0}{X_{\max}}(1 - e^{\mu_0 t})\right)$$

Here, Eq. (5) can be used to calculate approx. GSH concentration produced at any given time (t) in fermentation.

2.7.3 Glucose Consumption Kinetics

The modified Luedeking-Piret equation is used to describe the glucose consumption kinetics in yeast culture. This model consider that, the substrate utilized by cells mainly functions for biomass and product formation and maintenance activities in cell (Weiss and Ollis 1980).

$$\frac{dS}{dt} = -\frac{1}{Y_{X/S}}\frac{dX}{dt} - \frac{1}{Y_{P/S}}\frac{dP}{dt} - K_e X \tag{6}$$

Substituting $r_{fp} = -Y_{p/s}\, r_{fs}$ values in Eq. (6) and rearranging,

$$\frac{dS}{dt} = -\gamma\frac{dX}{dt} - \eta X \tag{7}$$

where r_{fp} is the rate of product formation, r_{fs} is the rate substrate utilization

$$\text{And}\quad \gamma(gS/gX) = \frac{1}{Y_{X/S}} + \frac{\alpha}{Y_{P/S}} \qquad \eta(gS/gX.h) = \frac{\beta}{Y_{P/S}} + K_e$$

Equation (7) is the modified Luedeking-Piret equation for substrate utilization kinetics. Rearranging Eq. (7)

$$-\frac{dS}{dt} = \gamma + \frac{\eta}{\mu} \tag{8}$$

Integrating Eq. (8) with initial conditions $X = X_0$ ($t = 0$) and $S = S_0$ ($t = o$) give

$$S_t = S_0 - \gamma m(t) - \eta n(t) \tag{9}$$

where

$$m(t) = \left[\frac{X_0 e^{\mu_0 t}}{1 - \left(\frac{X_0}{X_{max}}\right)(1 - e^{\mu_0 t})} - 1\right]$$

and

$$n(t) = \frac{X_{max}}{\mu_0}\ln\left[1 - \frac{X_0}{X_{max}}(1 - e^{\mu_0 t})\right]$$

This Eq. (9) can be used to obtain the residual substrate concentration in the production media.

3 Results and Discussion

3.1 Glutathione Production (Product Formation Kinetics)

The GSH production kinetics using *S. cerevisiae* NCIM 3345 was studied with the optimized medium and process parameters; temp at 30 °C, media pH 5.5 and mixing at 150 rpm. Figure 1, illustrates the kinetic study profiles of biomass formation, GSH production, and substrate consumption with respect to time using the optimized media and process parameters.

The rate of GSH production was found to be significantly increasing along with the exponential growth phase of micro-organisms proving that the GSH is being the growth associated product. The highest GSH concentration obtained was 157.5 mg/L in 16 h of the fermentation period. After 16th h, production reduces slowly. This might be due to the production of ethanol. GSH production and accumulation in yeast cells are favorable in lower ethanol concentration, while higher ethanol concentration inhibits the glutathione production. The maximum biomass concentration achieved was 3.48 g/L in 16 h duration as stationary phase hadn't contributed towards the increase in biomass concentration. First 4 h duration was an adaptation phase of cells to operating conditions and optimized production media. Next 6 h duration in fermentation was characterized as an exponential growth phase of the microorganism. The rate of product formation and substrate consumption were observed to be highest in this period. Yeast cells had consumed almost 92 % of the glucose till the fermentation ends and maximum GSH production was obtained at this stage correspondingly. The mathematical expression of obtained data from experimentation concluded that, unstructured kinetic models have the ability to explain and deliver process understating of the fermentation process for Glutathione production.

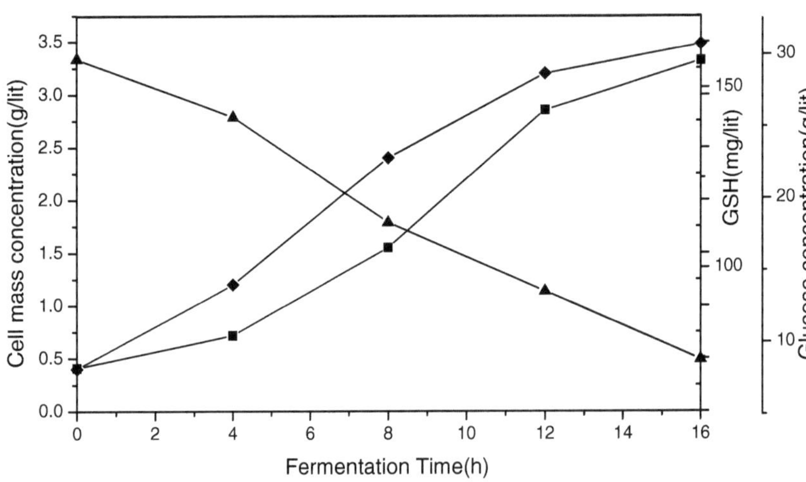

Fig. 1 Fermentation profile of cell mass concentration (*filled diamond*), glucose concentration (*filled triangle*) and GSH production (*filled square*)

3.2 Unstructured Kinetic Models

The models used in this study had provided a good estimation of predictive capability despite having very limited cellular mechanism information. Following Table 1, gives the calculated values for the estimated kinetic parameters used in unstructured models.

The relationship between actual versus predicted values is explained briefly using statistical tools such as regression coefficient (R^2) values for biomass growth, GSH production and glucose utilization profiles observed in fermentation. The regression coefficient measures the strength of linear relationship between the experimental (Actual) and predicted values obtained using the kinetic models. The linear relationship modeled by the straight line illustrates the steady increase or decrease. In Fig. 2a unbroken line shows estimated the response of Logistic model and broken line experimental response. Figure 2b gives a comparison of actual

Table 1 Estimated kinetic model parameters for GSH production

Parameter	Value
μ_0 (1/h)	0.37
X_{max} (g/L)	5.60
X_0 (g/L)	0.88
α (gP/gX)	8.95
β (gP/gX.h)	2.04
γ (gS/gX)	4.10
η (gS/gX.h)	0.52
$Y_{x/s}$ (g/g)	0.15
$Y_{p/s}$ (mg/g)	4.09
$Y_{p/x}$ (mg/g)	27.76
K_e (g/g.h)	0.018

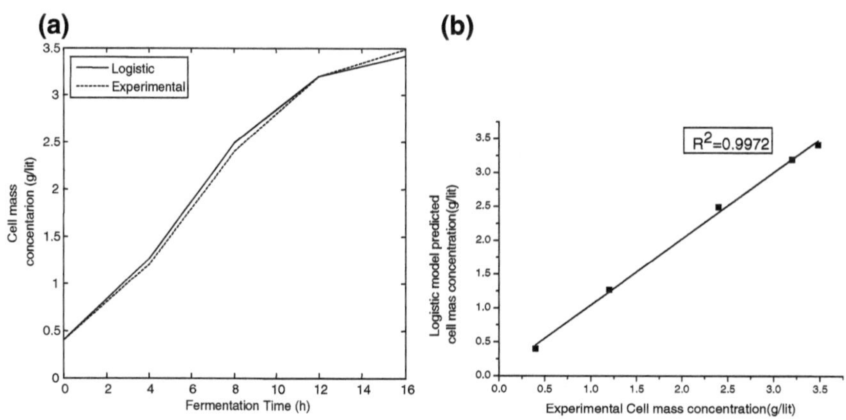

Fig. 2 **a** Experimental (*dashed line*) and model prediction (*dotted line*) biomass concentration. **b** Comparison of actual versus predicted biomass concentration by Logistic mode

versus predicted values obtained for biomass concentration with ($R^2 = 0.9972$). Figure 3a show response of Luedeking-Piret model for GSH production and Fig. 3b show comparison between experimental and model predicted GSH production with ($R^2 = 0.9814$). In Fig. 4a response of Modified Luedeking-Piret for glucose consumption and comparison of experimental and model predicted consumption shown in Fig. 4b. ($R^2 = 0.9535$). As observed in figures, product formation rate was maximum in the exponential growth phase of the cells i.e. ('α' ≫ than 'β') thus, the product glutathione was considered as primary metabolite due to its growth associated production during the fermentation process.

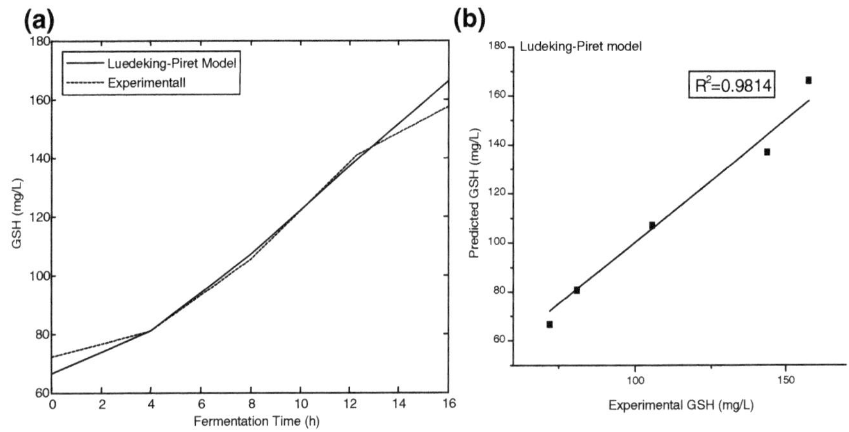

Fig. 3 **a** Experimental (*dashed line*) and model prediction (*dotted line*) of GSH. **b** Comparison of experimental and predicted GSH by Luedeking-Piret model

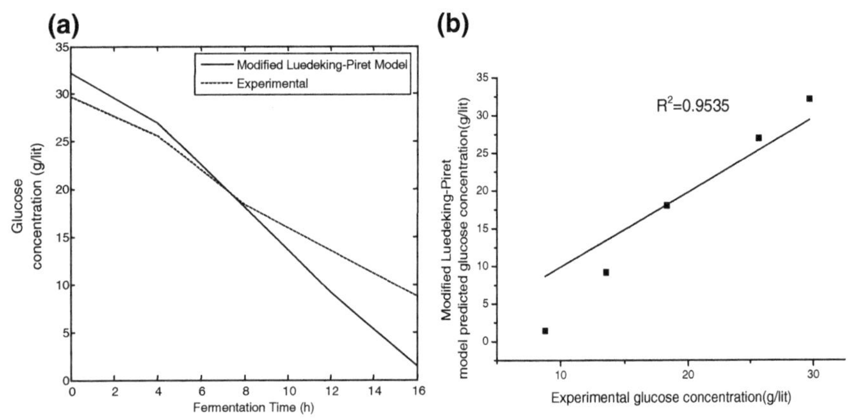

Fig. 4 **a** Experimental (*dashed line*) and model prediction (*dotted line*) Glucose consumption. **b** Comparison of experimental and predicted glucose consumption by modified Luedeking-Piret model

4 Conclusion

The mathematical modeling and expression of unstructured kinetics models were found to be relatively simple and easy. The greater prediction capability of the models had made them highly flexible for applying in any fermentation processes. In the present study, the models like Logistic equations for biomass growth, Luedeking-Piret model for GSH production and modified Luedeking-Piret model for glucose consumption i.e. substrate utilization were found best suitable to describe the glutathione production process using *Saccharomyces cerevisiae* NCIM 3345. All the models tested were able to predict profiles with higher R^2 values. The parameter of Luedeking-Piret model clearly indicated that GSH production is strictly growth associated.

Acknowledgments We are very much grateful that, department of biotechnology at KIT's College Of Engineering provided us an opportunity to utilize their facilities to conduct this research work as expected.

References

Alfafara, C.G., Kanda, A., Shioi, T., Shimizu, H., Shioya, S., Suga, K.: Effect of amino acids on glutathione production by *Saccharomyces cerevisiae*. Appl. Microbiol. Biotechnol. **36**, 538–540 (1992)

Berhane, K., Widersten, M., Engstrom, A., Kozarich, J.W., Mannervik, B.: Detoxication of base propenals and other, unsaturated aldehyde products of radical reactions and lipid peroxidation by human glutathione transferase. Proc. Natl. Acad. Sci. U S A. **91**, 1480–1484 (1994)

Cha, J.Y., Park, J.C., Jeon, B.S., Lee, Y.C., Cho, Y.S.: Optimal fermentation conditions for enhanced glutathione production by *Saccharomyces cerevisiae* FF-8. J. Microbiol. **42**, 51–55 (2004)

Chandrasekhar, K., Felse, P.A., Panda, T.: Optimization of temperature and initial pH and kinetic analysis of tartaric acid production by *Gluconobacter suboxydans*. Bioprocess. Eng. **20**, 203–207 (1999)

Cohn, V.H., Lyle, J.: A fluorometric assay for glutathione. Anal. Biochem. **14**, 434–440 (1966)

Ghaly, A.E., Kamal, M., Correia, L.R.: Kinetic modeling of continuously submerged fermentation of cheese whey for single cell protein production. Bioresour. Technol. **96**, 1143–1152 (2005)

Goto, S., Yoshida, K., Morikawa, T., Urata, Y., Suzuki, K., Kondo, T.: Augmentation of transport for cisplatin-glutathione adduct in cisplatin-resistant cancer cells. Cancer Res. **55**, 4297–4301 (1995)

Li, Y., Chen, J., Zhou, N., Fu, W., Ruan, W., Lun, S.: Effects of amino acids and yeast extract on glutathione production. Chin. J. Pharm. **29**, 537–542 (1998)

Miller, G.L.: Use of Dinitrosalicylic acid reagent for determination of reducing sugar. Anal. Chem. **31**, 426–428 (1959)

Pastore, A., Federici, G., Bertini, E., Piemonte, F.: Analysis of glutathione: implication in redox and detoxification. Clin. Cham. Acta. **33**, 19–39 (2003)

Perego, P., Weghe, J.V., Ow, D.W., Howell, D.W.: The role of determinants of cadmium sensitivity in the tolerance of *Schizosaccharomyces pombe* to cisplatin. Mol. Pharmacol. **51**, 12–18 (1997)

Sakato, K., Tanaka, H.: Advanced control of glutathione fermentation process. Biotechnol. Bioeng. **40**, 904–912 (1992)

Sollod, C.C., Jenns, A.E., Daub, M.E.: Cell surface redox potential as a mechanism of defense against photosensitizers in fungi. Appl. Environ. Microbiol. **58**, 444–449 (1992)

Wei, G., Li, Y., Chen, J.: Effect of surfactants on the extracellular accumulation of glutathione by *Saccharomyces cerevisiae*. Process Biochem. **38**, 1133–1138 (2003a)

Wei, G., Li, Y., Chen, J.: Application of a two-stage temperature control strategy for enhanced glutathione production in the batch fermentation by *Candida utilis*. Biotechnol. Lett. **25**, 887–890 (2003b)

Weiss, R.M., Ollis, D.F.: Extracellular microbial polysaccharides I. Substrate, biomass and product kinetic equations for batch Xanthan gum fermentation. Biotechnol. Bioeng. **22**, 859–873 (1980)

Statistical Optimization of Lactic Acid Extraction from Fermentation Broth Using Emulsion Liquid Membrane

Avinash Thakur, Parmjit Singh Panesar and Manohar Singh Saini

1 Introduction

Lactic acid ($CH_3CHOHCOOH$) first refined in 1780 by the Swedish chemist Scheele from sour milk is utilized in the manufacturing of a wide range of industrial compounds as acidulant, flavor and preservative for food, textile, pharmaceuticals, cosmetics and chemical industries etc. and a variety of industrially important chemicals such as propylene oxide, propylene glycol, 2, 3-pentanedione and lactate esters can be easily produced (Se-Kwon et al. 2012). It is also polymerized to produce highly demanded biodegradable poly-lactic acid (PLA) (Mohamed et al. 2011; Pattana et al. 2010).

Lactic acid can be obtained either by chemical synthesis or by fermentation. The chemical synthesis route gives only a racemic mixture of L- and D-lactic acid from petrochemical resources (a major cost-contributor). While stereoisomer of lactic acid can be produced by a fermentation process from renewable resources by

A. Thakur (✉)
Department of Chemical Engineering, Sant Longowal Institute of Engineering and Technology, Longowal, Sangrur 148106, Punjab, India
e-mail: avin_thakur2@yahoo.com

P.S. Panesar
Biotechnology Research Laboratory, Department of Food Engineering and Technology, Sant Longowal Institute of Engineering and Technology, Longowal, Sangrur 148106, Punjab, India
e-mail: pspanesarrr@yahoo.com

M.S. Saini
Guru Nanak Dev Engineering College, Ludhiana, Punjab, India
e-mail: mssaini58@hotmail.com

Prasanna B.D. et al. (eds.), *Biotechnology and Biochemical Engineering*,
DOI 10.1007/978-981-10-1920-3_3

choosing the appropriate microorganism and offers several other advantages over chemical synthesis such as, low cost of substrates, a low production temperature and low energy consumption (Niju et al. 2004). Worldwide about 90 % of lactic acid is obtained by fermentation route (Joglekar 2006). Both bacterial and fungal strains can be utilized for lactic acid production. But normally fungal strains grow very slowly and exhibits relatively lower productivity (Se-Kwon et al. 2012).

For the manufacturing of lactic acid by fermentation, inexpensive renewable raw materials, such as molasses, starch (corn starch, wheat starch, potato starch) and lignocellulose (corn cobs and woody materials) have been utilized, tested or proposed by various investigators. Cane molasses (approximately 45–60 % (w/w) total sugars), is an excellent source of carbohydrates along with other essential nutrients and stimulates cell growth (Ke and Ping 2014; Umar et al. 2012; Dan et al. 2010).

Cell immobilization is an ideal approach for getting improved fermentation performance since immobilized cells have numerous advantages over free cell (Mrinal et al. 2014). Due to the simplicity and non-toxic character, entrapment in calcium/sodium alginate gel has been frequently used immobilization method (Min-Tian et al. 2009; Farshid et al. 2011).

The recovery or separation of lactic acid from the fermentation broth is very complex and cumbersome process since other impurities such as residual sugar, impurity protein, pigment, mycelium and carbohydrates also coexists along with the produced acid (Tayyba et al. 2014; Xiaolin et al. 2012). The possibility to reduce cost also causes interest in lactic acid removal (Lucas 2013). Conventional purification technique consists of a numerous downstream treatment steps which are not environmental benign and economically favorable (Dey and Pal 2012; Xiaolin et al. 2012). An economic separation method for the lactic acid recovery from fermentation broth is required to compete with a synthetic process (Sushil et al. 2011).

Emulsion liquid membrane (ELM) technology, initially proposed by Li in 1968 (Jia et al. 2013) is a simplified extraction process with high extraction efficiency which combines both extraction and stripping stage to undertake simultaneously purification and concentration. ELM processes are simple, consume relatively low energy compared to other separation processes, provides high selectivity and enables very high and rapid mass transfer rate with a high degree of effectiveness as it offers large surface area per unit volume (Jia et al. 2013; Goyal et al. 2011). The enhanced selectivity can be obtained by addition of a carrier/extractant to the membrane phase (Julio et al. 2010). ELM could be cheaper (≤40 %) in comparison to other solvent extraction methods (Mokhtari and Pourabdollah 2012). ELM has been successfully employed by the various researchers for the separation of sugars, organic acids, proteins, amino acids and antibiotics (Julio et al. 2010).

Keeping in view the various advantages offered by ELM, its performance for lactic acid extraction from diluted and centrifuged fermentation broth has been studied and response surface methodology, a collection of statistical and mathematical techniques is useful for developing, improving and optimizing processes and was employed for process optimization of the process variables for maximizing the lactic acid extraction from diluted and centrifuged fermentation broth.

2 Materials and Methods

2.1 *Lactic Acid Production by Fermentation*

2.1.1 Procurement of Microbial Cultures and Cell Culture Preparation

The microbial strain *Lb. casei* MTCC 1423 utilized in present study was procured from Microbial Type Culture Collection (MTCC), Institute of Microbial Technology (IMTECH), Chandigarh, India. Freeze-dried microbes were cultured for 20 h at 37 °C (1 %, w/v) in sterile MRS (de Mann Rogosa Sharpe) broth. The obtained culture was then sub-cultured (37 °C, 20 h) twice in sterile MRS broth using 1 % (w/v) of inoculums for activation and adaptation.

2.1.2 Molasses, Corn Steep Liquor and Chemicals

Sugarcane molasses, the byproduct of cane sugar plants, was procured from Bhagwanpura Sugar Mill Limited Dhuri, Punjab, India. Corn steep liquor (CSL), waste water was procured from Sukhjeet Industries, Phagwara, Punjab, India. Sugarcane molasses and corn steep liquor were stored at 4 °C and were used without pretreatment. All the chemicals used during present study for experimental investigations were of analytical grade (HPLC grade for HPLC analysis).

2.1.3 Fermentation Media and Immobilization of Cells

Fermentation medium (50 mL) having molasses sugar (186.5 gm/L), $MnSO_4$ (20 mg/L), $CaCO_3$ (25 %, w/w) with respect to sugar content and CSL (2.5 %, v/v) in Erlenmeyer flasks was sterilized (121 °C, 15 psi for 20 min) before the fermentation. The procedure adopted by Ani et al. (2006) and Imdad et al. (2014) was used for the entrapment of *Lb. casei* MTCC 1423 cells. The bead coating was employed as described by Klinkenberg et al. (2010).

2.2 *Emulsion Liquid Membrane*

2.2.1 Membrane Preparation

For preliminary stability analysis and lactic acid extraction efficiency of ELM (w/o/w) in the presence of molasses in the external phase, the emulsion was obtained by mixing the internal stripping phase with membrane (organic) phase. The membrane phase contained 4.5 % (v/v) Span 80 as stabilizer, 0–10 % (v/v) tri-n- octyl amine (TOA), as carrier in n-hexane and 2 % (v/v) cyclohexanone (Prasad et al. 2010) (to reduce water co-transportation) was stirred at 200 rpm for 2 min using a magnetic stirrer. To this homogeneous membrane phase, 0.1 M stripping phase (Na_2CO_3 solution) with 1:1 (v/v) internal to organic phase ratio was added drop wise and was stirred at 2000 rpm using a four blade impeller stirrer (Model: IKA RW 20, Digital Dual Range Mixer from Cole-Parmer, India) for 20 min at room temperature (25 ± 2 °C) to form a stable liquid emulsion membrane.

2.2.2 Emulsion Stability and Lactic Acid Extraction with Molasses in External Phase

For stability analysis, the formed emulsions were poured into 100 mL of diluted centrifuged molasses solution in a 250 mL beaker (treat ratio = 2) at a low stirring speed. The molasses was diluted and centrifuged at a 7000 rpm for 10 min to make the conditions similar to that of after the fermentation of molasses by immobilized cells. The external phase pH was measured after different interval of experimental contact time. Synthetic lactic acid (0.05 M) was added in the diluted centrifuged molasses solution to check $\eta_{ext.}$ with 0.2 M Na_2CO_3 in internal phase of ELM.

2.3 *Lactic Acid Extraction from Fermentation Broth*

Keeping in view the stability of the emulsion in molasses solutions, an optimized formulation of water in oil emulsion obtained previous experimentation for maximizing the lactic acid extraction efficiency from aqueous lactic acid solution with slight modification was used for batch extraction of lactic acid after fermentation by double coated (first with chitosan and then by sodium alginate) immobilized *Lactobacillus (Lb.) casei* MTCC 1423 cells. The membrane phase was consisted of span-80 concentration (4.5 %, v/v), TOA (4.75 %, v/v) and 2 % (v/v) cyclohexanone dissolved in n-hexane. The water in oil emulsion was prepared by mixing of membrane phase (20 mL) with 0.2 M Na_2CO_3 (20 mL) at 2000 rpm for 20 min. For the extraction experiments, the fermentation broth containing lactic acid produced by *Lb. casei* MTCC 1423 cells was centrifuged at 6700 rpm for 10 min and analyzed for lactic acid concentration after adding the concentrated sulphuric acid to release the lactic acid and then diluted to obtain the desired concentration of lactic acid (0.06 M). The extraction was carried out at 200 rpm for 13.5 min with a treat ratio of 2 (v/v).

2.4 *Estimation of Total Sugar, Lactic Acid Concentration and* pH

The total sugar concentration was determined according to the phenol sulfuric acid method (Dubois et al.1996). The lactic acid concentration measurement was carried out by the HPLC method (Ginjupalli et al. 2013) using Shimadzu LC 2010 CHT (Shimadzu Corporation, Kyoto, Japan) equipped with low pressure quaternary gradient pump, dual wavelength UV-Visible detector and column oven. pH was measured using Eutech pH 5+.

2.5 *Calculation of Emulsion Breakage (%)*

The emulsion breakage (%) λ (percentage of stripping phase reagent leaked into the external phase) was calculated using pH method as (Hongpu et al. 2013)

$$\text{Emulsion breakage}(\%), \lambda = \frac{M_s V_E (10^{pH-14} - 10^{pH_0-14})}{c_{s0} V_s} \times 100 \quad (1)$$

where M_s is the molar mass of stripping reagent, pH_0 and pH is initial pH of external phase and after certain contact time respectively. c_{s0} is initial stripping reagent concentration in stripping phase.

2.6 *Calculation of Lactic Acid Extraction*

The extraction efficiency had been calculated using the equation:

$$\text{Extraction efficiency}, \eta_{ext} = \frac{C_{l0} - C_{l\tau}}{C_{l0}} \times 100 \quad (2)$$

where C_{l0} and $C_{l\tau}$ is the lactic acid concentration in external phase initially at time t = 0, and after contacting the aqueous phase with emulsion liquid membrane for desired time respectively.

2.7 *Optimization of Lactic Acid Extraction from Fermentation Broth*

2.7.1 Experimental Design

For the optimization of extraction efficiency, the experiments were conducted according to experimental design obtained from central composite rotatable design (CCRD) with three variables viz. span 80 concentration, c_s (v/v), stirring speed, ω (rpm) and batch extraction time (τ, min) on the extraction efficiency (η_{ext}) at five levels each using Design-Expert 7.16 software (Statease Inc., Minneapolis, USA) at the ranges as shown in Table 1. The experimental plan in un-coded form of process variables along with experimental results is as shown in Table 2.

Table 1 Range of different variables for lactic acid extraction from fermentation broth using ELM

	Factors	Process parameters	Level				
		Coded values	$-\alpha$	−1.000	0.000	+1.000	$+\alpha$
Un-coded values	X_1	Span 80 concentration, c_s (%, v/v)	4	4.5	5.25	6	6.5
	X_2	Carrier concentration, ψ (%, v/v)	3.25	5	7.5	10	11.75
	X_3	Batch extraction time, τ (min)	10	15	22.5	30	35

Table 2 Experimental design using central composite rotatable design for lactic acid extraction from fermentation broth by emulsion liquid membrane

Span 80 concentration c_s (%, v/v)	Carrier concentration ψ (%, v/v)	Batch extraction time τ (min)	Extraction efficiency η_{ext} (%)
6	10	15	58.56
5.25	7.5	22.5	76.54
4.5	10	15	62.68
5.25	7.5	10	59.42
6.5	7.5	22.5	61.26
6	10	30	82.74
4.5	5	30	67.2
4.5	5	15	65.14
5.25	7.5	35	81.25
5.25	3.25	22.5	57.21
5.25	7.5	22.5	76.84
4	7.5	22.5	68.24
6	5	15	51.5
5.25	7.5	22.5	76.14
5.25	7.5	22.5	76.65
6	5	30	63.11
4.5	10	30	77.23
5.25	11.75	22.5	72.43
5.25	7.5	22.5	75.12

2.7.2 Experimental Procedure

The external phase utilized during the optimization experiments and the ELM employed was obtained as per procedure described above (2.3) in combination process variables values of Table 2.

2.7.3 Statistical Analysis and Optimization

Response surface methodology was applied to the experimental data using Design-Expert 7.16 software for the generation of response surface and contour plots showing the interactions of any two independent variables while keeping the values of the other variables as constant. The same software was used for the optimization of process variables.

3 Results and Discussion

3.1 Effect of Loading External Phase with Molasses on Emulsion Stability and Extraction Efficiency

The increase in the molasses content in the external phase had caused a decrease in the emulsion stability (Fig. 1). Certain components in the molasses might be destabilizing the emulsion either by interacting with surfactant Span 80 or by altering its properties (Julio et al. 2010). Moreover with the decrease in the dilution rate, the viscosity of the external phase also increases. It may have promoted the shear force phenomenon acting on the emulsion at 200 rpm stirring speed. Emulsion had been dispersed in the external phase containing synthetic lactic acid and centrifuged diluted molasses by varying the dilution factor, D(4–10). The extraction efficiency had been observed to be increased with the increase in dilution of the molasses (Fig. 2).

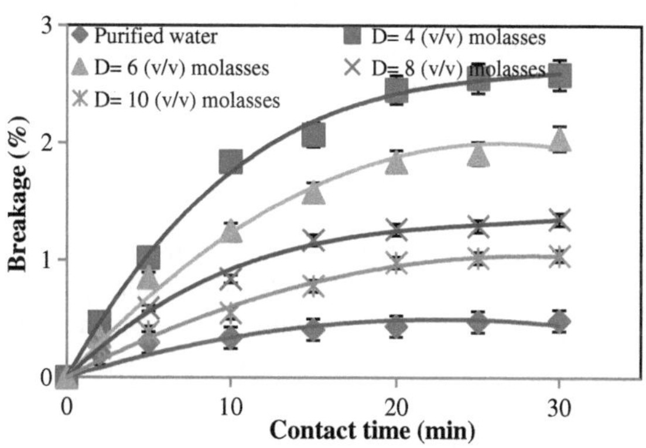

Fig. 1 Effect of loading external phase with molasses on emulsion stability

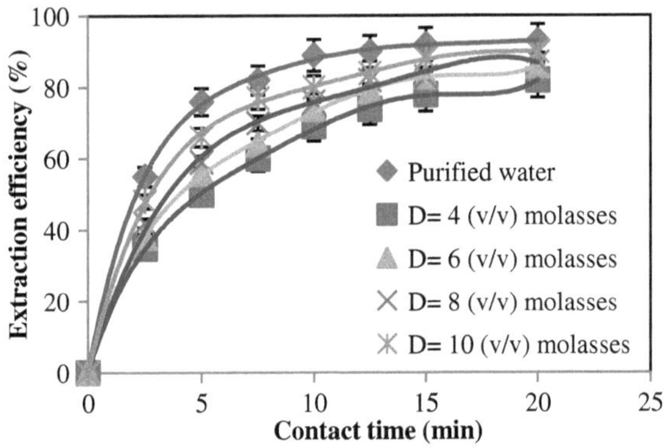

Fig. 2 Effect of loading external phase with molasses on lactic acid extraction by ELM

3.2 *Lactic Acid Extraction Efficiency from Fermentation Broth Using Emulsion Liquid Membrane*

The lactic acid extraction using ELM from the diluted and centrifuged fermentation broth by immobilized *Lb. casei* MTCC 1423 cells after the 60 h of incubation time at incubation temperature of 37.5 °C with optimized conditions was carried out. The extraction efficiency (82 %) after 13.5 min of the batch extraction time was obtained. During the extended batch extraction time, the stability of the emulsion tends to slightly decreased with 4.5 % (v/v) span 80 concentration. This could be due to interaction of certain ingredients of the fermentation broth with the surfactant (Julio et al. 2010).

3.3 *Optimization of Lactic Acid Extraction from Fermentation Broth Using Emulsion Liquid Membrane*

3.3.1 Regression Model

For extraction efficiency (%), the fit summary concluded that the quadratic model is significant for analysis. The ANOVA for the quadratic model before elimination of non-significant terms had nine terms which contain three linear terms, three quadratic terms and three two-factorial interactions. Probability $p > f$ values were employed as a tool to verify the significance of each of the coefficients. The ANOVA table for a quadratic model after backward elimination with alpha out = 0.0500 has been shown in Table 3.

The model results indicated that model is significant (R^2 and adjusted R^2 are 98.79 and 98.48 % respectively), lack of fit has been found to be non-significant.

Table 3 Regression model and ANOVA for lactic acid extraction using emulsion liquid membrane from fermenation broth

Source	Sum of squares	Degree of freedom	Mean square	*f*- value	$p > f$
Model	1480.65	9	164.51	706.75	<0.0001
Span 80 concentration (X_1)	57.73	1	57.73	248.01	<0.0001
Trioctyl amine (TOA) concentration (X_2)	262.34	1	262.34	1127.03	<0.0001
Batch extraction time (X_3)	581.48	1	581.482	2498.03	<0.0001
$(X_1) \times (X_2)$	45.69	1	45.6968	196.31	<0.0001
$(X_1) \times (X_3)$	45.985	1	45.98	197.54	<0.0001
$(X_2) \times (X_3)$	78.50	1	78.500	337.23	<0.0001
$(X_1) \times (X_1)$	226.71	1	226.71	973.94	<0.0001
$(X_2) \times (X_2)$	223.96	1	223.96	962.14	<0.0001
$(X_3) \times (X_3)$	60.24	1	60.24	258.79	<0.0001
Residual	2.09	9	0.232		
Lack of fit	0.21	5	0.042	0.091	0.9892*
Pure error	1.88	4	0.47		
Corrected total	1482.74	18			
Standard Deviation = 0.69	R^2 = 0.9879		Adjusted R^2 = 0.9848		
Mean = 69.09	Predicted R^2 = 0.9812		Adequate precision = 64.122		
PRESS = 6.81	C.V. = 0.99				

*non-significant at 5 % level

The final response equation for extraction efficiency (%) is given as follows (in terms of coded factors)

$$\begin{aligned}\text{Extraction efficiency}, \eta_{ext}(\%) = {} & 76.26 - 2.05X_1 + 4.38X_2 + 6.26X_3 \\ & + 2.39X_1X_2 + 2.42X_1X_3 + 3.13X_2X_3 \\ & - 4.17X_1X_1 - 4.05X_2X_2 - 2.10X_3X_3 \end{aligned} \quad (3)$$

Response surface curves for lactic acid from centrifuged and diluted (D = 10) fermentation broth by ELM using TOA are shown in Figs. 3, 4, 5. The interactive effect of carrier (TOA) and surfactant (span 80) concentration had been depicted in the Fig. 3. It is evident that with the increase in span 80 concentration the extraction efficiency had decreased. The decrease in η_{ext} is significant at the higher carrier concentration. This might be due to the reason that both the variables had contributed towards the growth of membrane phase-internal phase interfacial viscosity resulting an increase in resistance to the diffusion of lactic acid across it (Hongpu et al. 2013). The extraction efficiency had increased with the increase in carrier concentration, but tends to slightly decrease at the higher concentration.

The extraction efficiency had increased with the increase in batch extraction time (Fig. 4). The extraction efficiency had achieved an approximately early steady state

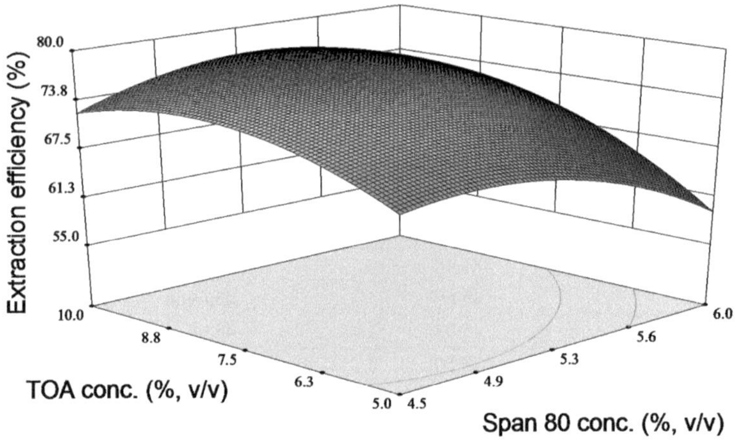

Fig. 3 Effect of carrier concentration and span 80 concentration on extraction efficiency of ELM from fermentation broth

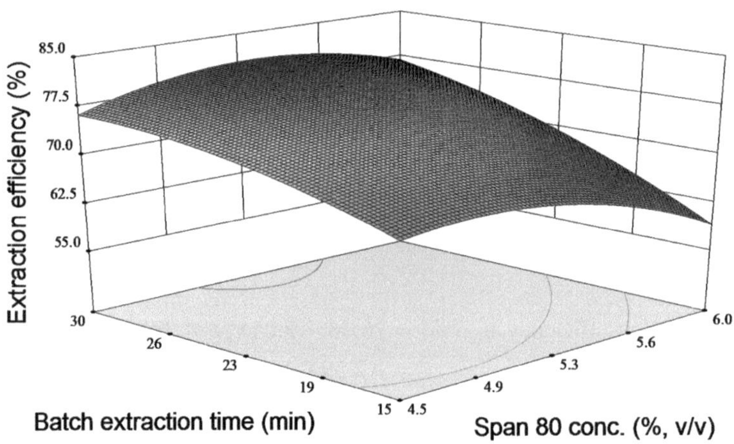

Fig. 4 Effect of batch extraction time and span 80 concentration on extraction efficiency of ELM from fermentation broth

at lower surfactant concentration while at higher concentration it was taking a longer time to achieve it. This might be owing to the fact that at higher surfactant concentration globule in the external phase (fermentation broth) did not disperse very well and the surface area for mass transfer has been significantly decreased. At the same time the surfactant hydration also significantly increases the co-transport of water molecules thus diluting the internal phase (Raja et al. 2014).

The extraction efficiency had increased with the increase in TOA concentration and batch extraction time (Fig. 5). The extraction time required to attain the maximum extraction at higher carrier concentration is more in comparison to that of

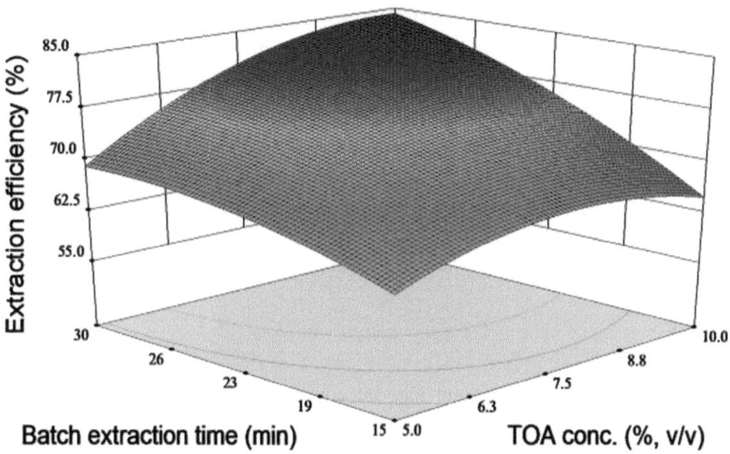

Fig. 5 Effect of batch extraction time and carrier concentration on extraction efficiency of ELM from fermentation broth

from the aqueous phase. There might be due to chemical complexity arisen due to certain other components present in the fermentation broth and caused a decrease in the lactic acid transport rate through the membrane phase (Julio et al. 2010).

3.4 *Optimization of Lactic Acid Extraction from Fermentation Broth*

A numerical optimization technique was employed to obtain the workable optimum conditions for the extraction of lactic acid from the centrifuged and diluted fermentation broth. The optimum conditions obtained for maximum lactic acid extraction (82.67 %) were: span 80 concentration, c_s: 5 % (v/v); TOA concentration, ψ: 9.5 % (v/v); batch extraction time τ: 28 min. A set of experiments were carried to validate the results which were predicted by the model for the maximum extraction efficiency using predicted optimum values of variables. The maximum value of extraction efficiency achieved from fermentation broths was around 79.87 %.

4 Conclusion

L(+)-lactic acid extraction using ELM from the diluted and centrifuged fermentation broth by immobilized *Lb. casei* MTCC 1423 cells after the 60 h of incubation time with optimized conditions was carried out. The extraction efficiency (82 %)

after 13.5 min of the batch extraction time was obtained. Keeping in view the stability aspect using ELM for lactic acid extraction from fermentation broth, further investigations had been carried out with different combination of surfactant concentration along with by varying the extractant concentration and batch extraction time using response surface methodology. The optimum conditions obtained for maximum L(+)-lactic acid extraction (82.67 %) were: span 80 concentration 5 % (v/v); TOA concentration 9.5 % (v/v); batch extraction time 28 min.

References

Ani, I., Wahidin, S.: Effect of sodium alginate concentration, bead diameter, initial pH and temperature on lactic acid production from pineapple waste using immobilized *Lactobacillus delbrueckii*. Process Biochem. **41**, 1117–1123 (2006)

Dan, C.V., Adriana, P., Fancisc, V.D., Carmen, S.: HPLC characterization of lactic acid formation and ftir fingerprint of probiotic bacteria during fermentation processes. Not. Bot. Horti. Agrobo. **38**(1), 109–113 (2010)

Dey, P., Pal, P.: Direct production of L(+) lactic acid in a continuous and fully membrane-integrated hybrid reactor system under non-neutralizing conditions. J. Membr. Sci. **389**, 355–362 (2012)

Dubois, M., Gilles, K.A., Hamilton, J.K., Robers, P.A., Smith, F.: Colorimetric method for determination of sugar and related substances. Anal. Chem. **28**, 350–356 (1996)

Farshid, G., Habibollah, Y., Abbas, E.S., Ghasem, N.: Cane molasses fermentation for continuous ethanol production in an immobilized cells reactor by *Saccharomyces cerevisiae*. Renew. Ener. **36**, 503–509 (2011)

Ginjupalli, K., Armugam, K., Shavi, V.G., Averineni, R.K., Mahalingam, B., Udupa, N.: Development of RP-HPLC method for simultaneous estimation of lactic acid and glycolic acid. Der Pharma Chemica. **5**(4), 335–340 (2013)

Goyal, R.K., Jayakumar, N.S., Hashim, M.A.: Chromium removal by emulsion liquid membrane using [BMIM]+[NTf2]—as stabilizer and TOMAC as extractant. Desalination **278**, 50–56 (2011)

Hongpu, J., Wei, P., Juntong, Z., Chunjian, X.: Extraction performance of bisphenol A from aqueous solutions by emulsion liquid membrane using response surface methodology. Desalination **313**, 36–43 (2013)

Imdad, K., Huang, S., Bo, L., Bin, W., Ra, Aamir, Chun, L.: Efficient biosynthesis of glycyrrhetic acid 3-O-mono-β-D-glucuronide(GAMG) in water-miscible ionic liquid by immobilized whole cells of *Penicillium purpurogenum* Li-3 in alginate gel. Chem. Eng. Sci. **106**, 136–143 (2014)

Jia, O., Rui, M., Zhaojuan, Z.C.C., Min, Z., Ting, J.: Open fermentative production of L-lactic acid by *Bacillus sp*. strain NL01 using lignocellulosic hydrolyzates as low-cost raw material. Bioresour. Technol. **135**, 475–480 (2013)

Joglekar, H.G., Imran, R., Suresh, B., Kulkarni, B.D., Ajit, J.: Review-comparative assessment of downstream processing options for lactic acid. Sep. Purif. Technol. **52**, 1–17 (2006)

Julio, B., Leo, D.P., Geran, A.: Gibberellic acid extraction from aqueous solutions and fermentation broths by using emulsion liquid membranes. J. Membr. Sci. **348**, 91–98 (2010)

Ke, X., Ping, X.: Efficient production of L-lactic acid using co-feeding strategy based on cane molasses/glucose carbon sources. Bioresour. Technol. **153**, 23–29 (2014)

Klinkenberg, G., Lystad, K.Q., Levine, D.W., Dyrset, N.: Cell release from alginate immobilized *Lactococcus lactis ssp. lactis* in chitosan and alginate coated beads. J. Dairy Sci. **84**, 1118–1127 (2010)

Lucas, D., Phylipe, A. C., Julio, L. d. S. J., Leonardo, H. d. O., Martin, A.: Liquid equilibrium data for ternary systems of water + lactic acid + C_4-C_7 alcohols at 298.2 K and atmospheric pressure, Fliuid Phase Equilibr. **354**, 12–18 (2013)

Min-Tian, G., Takashi, S., Nobuhiro, I., Haruo, T.: Fermentative lactic acid production with a metabolically engineered *Yeast* immobilized in photo-crosslinkable resins. Biochem. Eng. J. **47**, 66–70 (2009)

Mohamed, A.Abdel-R, Yukihiro, T., Kenji, S.: Lactic acid production from lignocellulose-derived sugars using lactic acid bacteria, overview and limits. J. Biotechnol. **156**, 286–301 (2011)

Mokhtari, B., Pourabdollah, K.: Inclusion desalination of alkali metal cations by emulsion liquid membranes and nano-baskets of p-tert-calix105arene bearing di-[N-(X)sulfonylcarboxamide] and di-(1-propoxy) in para-cone conformation. Desalination, **292**, 1–8 (2012)

Mrinal, N.K., Angelika-Ioanna, G., Jean Bernard, M., Panagiotis, K., Argyro, B., Athanasios, A. K., Maria, K.: Lactic acid fermentation by cells immobilised on various porous cellulosic materials and their alginate/poly-lactic acid composites. Bioresour. Technol. **165**, 332–335 (2014)

Niju, N., Pradip, K.R., Aradhana, S.: L(+) lactic acid fermentation and its product polymerization. Electron. J. Biotechn. **7**(2), 167–179 (2004)

Pattana, L., Arthit, T., Vichean, L., Lakkana.: L. Acid hydrolysis of sugarcane bagasse for lactic acid production. Bioresour. Technol. **101**, 1036–1043 (2010)

Prasad, S.K., Shrikant, J.W., Vijaykumar, V.M.: Process intensification in extraction by liquid emulsion membrane (LEM) process, a case study; enrichment of ruthenium from lean aqueous solution. Chem. Eng. Process. **49**, 441–448 (2010)

Raja, N.R.S., Norasikin, O., Nor, A.S.A.: Emulsion liquid membrane stability in the extraction of ionized nanosilver from wash water. J. Ind. Eng. Chem. **20**, 3243–3250 (2014)

Se-Kwon, M., Young-Jung, W., Gi-Wook, C.: A novel lactic acid bacterium for the production of high purity L-lactic acid, *Lactobacillus paracasei subsp. paracasei* CHB2121. J. Biosci. Bioeng. **114**, 155–159 (2012)

Sushil, K., Dipaloy, D., Babu, B.V.: Estimation of equilibrium parameters using differential evolution in reactive extraction of propionic acid by tri-n-butyl phosphate. Chem. Eng. Process. **50**, 614–622 (2011)

Tayyba, G., Muhammad, I., Zahid, A., Tahir, A., Zubia, Z., Asma, T., Muhammad, K., Nudrat, E., Sajid, M.: Recent trends in lactic acid biotechnology, a brief review on production to purification. J. Rad. Res. Ap. Ac. **7**, 222–229 (2014)

Umar, F., Faqir, M.A., Tahir, Z., Sajjad-U-R., Muhammd, A.R., Anwaar, A., Kashif, A.: Optimization of lactic acid production from cheap raw material, sugarcane molasses. Pak. J. Bot. **44**(1), 333–338 (2012)

Xiaolin, W., Yaoming, W., Xu, Z., Tongwen, X.: In situ combination of fermentation and electrodialysis with bipolar membranes for the production of lactic acid, operational compatibility and uniformity. Bioresour. Technol. **125**, 165–171 (2012)

Optimization of Microwave Assisted Extraction of Pectin from *Helianthus annuus* Head Using Response Surface Methodology

B.K. Aarthi, V. Aswini, M. Lakshmi Priya, M. Nirosha and M. Shanmugaprakash

1 Introduction

Pectin is a heteropolysaccharide consisting mainly of 65 % galacturonic acid units linked to neutral sugars by rhamnose unit to form branched chain contained in a middle lamella and primary cell wall of terrestrial plants. Its application is wide in food industry as gelling agent and pharmaceutical industry as drug carriers (Liu et al. 2003). Source of pectin includes citrus peel, gooseberries, apple pomace , oranges, citrus fruits which contains higher pectin. Citrus peel contains relatively higher percent of pectin that is 25 % compared to apple pomace (15–18 %). Alternate sources for pectin is sugar beet waste (Morris et al. 2010) and mango waste.

Waste resources like sunflower heads that are thrown waste can be utilized for the extraction of pectin efficiently. Sunflower heads are rich in Lignin (12.3 %), D-Glucose (23.5 %), low methoxy (LM) pectin and several other biomolecues and the entire plant contributes to about 15–24 % of pectin. Commercial extraction of pectin is done under high temperature using acids i.e., hot acid extraction (Srivastava et al. 2011). Considering the process economics and time consumption of hot acid method, there are many alternate methods of extraction of pectin like Microwave assisted extraction, Ultrasound assisted extraction (Li et al. 2014), using acid solvents have been done by researchers all around the world. Microwave assisted extraction is an efficient method that reduces the time and energy for extraction and gives higher yield (Fishman et al. 2006; Kratchanova et al. 2004).

B.K. Aarthi · V. Aswini · M. Lakshmi Priya · M. Nirosha · M. Shanmugaprakash (✉)
Downstream Processing Laboratory, Department of Biotechnology, Kumaraguru College of Technology, Coimbatore 641049, Tamil Nadu, India
e-mail: sunbioin@gmail.com

Prasanna B.D. et al. (eds.), *Biotechnology and Biochemical Engineering*,
DOI 10.1007/978-981-10-1920-3_4

Several authors reported that various factors such as time, microwave power, solid-liquid ratio and pH affecting the efficiency of extraction (Koh et al. 2014; Rahmati 2015; Kute et al. 2015). If the multiple factors and their interaction affects a desired output, response surface methodology (RSM) is an effective tool to optimize the process (Nwabueze 2010). Over a decade, RSM has been effectively adopted by several researchers for their various work (Prakash Maran et al. 2013; Shanmugaprakash et al. 2014, 2015).

Among that, BBD is one of the response surface methodologies was used for optimization of parameters since BBD limits the number of experiments for extraction. The objective of the current work is to optimize the process parameters for the extraction of pectin from sunflower heads by Box-Behnken design and to characterize the extracted pectin using Fourier transform infrared spectroscopy.

2 Materials and Methods

2.1 *Raw Materials and Reagents*

Raw materials (sunflower heads) were collected from sunflower garden, Coimbatore, Tamil Nadu. Sunflowers heads were shade dried powdered. The powdered sample was used for further analysis. Citric acid (SDFCL, Mumbai) was used to dissolve the sample powder. Iso-propanol (SDFCL, Mumbai) was used to precipitate pectin (Yoo et al. 2012).

2.2 *Extraction of Pectin Using Microwave Assisted Radiation*

According to the method described by Wang et al. (2007), extraction process was performed with a household microwave oven (LG, 3850 W) with adjustable irradiation time and microwave power under different conditions (Table 1). About 1 g of sample powder was weighed and dissolved in citric acid in a glass beaker and the beaker was placed in the rotating disk exposed to microwave radiation at different time for selected microwave power, temperatures and solid-liquid ratios. After that, the mixture was then cool down to room temperature, centrifuged (10,000 rpm at 4 °C for 10 min). The mixture was filtered using Whatman no-1 filter paper and precipitation of supernatant was carried out with an equal volume of Isopropanol. The coagulated pectin was centrifuged (10,000 rpm at 4 °C for 10 min), and pellet was dried and dry weight was taken. The equation proposed by Maran et al. (2014) was used to calculate pectin yield.

$$\mathrm{PY} = (m_0/m) \times 100 \quad (1)$$

where, m_0 (g) is the dry weight of pectin and m (g) is the dry weight of sunflower husk.

2.3 Experimental Design

In this work, RSM based BBD was chosen to find relationship of process variables on the pectin extraction. Four factors of variables such as pH (X_1), microwave power (X_2), time (X_3) and solid-liquid ratio (X_4) is shown (Table 1). The extracted pectin yield (Y) was selected as experimental response. Second order polynomial equation was derived, as shown in the following equation

$$\mathrm{Y} = \beta_0 + \sum_{j=1}^{K} \beta_i \mathrm{X}_j + \sum_{j=1}^{K} \beta_{jj} \mathrm{X}_j^2 + \sum_{i} \sum_{<j=2}^{K} \beta_{ij} \mathrm{X}_i \mathrm{X}_j + e_i \quad (2)$$

where, Y is the response; X_i and X_j represents variables (i and j range from 1 to k); β_0 is the model intercept coefficient; β_j, β_{jj} and β_{ij} represents interactive coefficients of linear, quadratic equation with second-order terms, respectively; k is the number of independent parameters (k = 4); and e_i is the error.

3 Results and Discussion

3.1 Modeling and Optimization of Extraction Process

Based on BBD design, experiments were performed in order to study the combined effects of the independent variables such as pH (X_1), microwave power (X_2), time (X_3) and solid-liquid ratio (X_4) on pectin extraction and the measured responses are listed in Table 2. The second order polynomial equation was generated which denotes the pectin extraction process, having higher *F*-value with low *p*-value.

Pectin extraction = −47.8233 + 71.09877 * pH + 0.200452 * S/L + 0.210228 * time + 0.010747 * micro oven − 0.20963 * pH * S/L − 0.00908 * pH * time + 0.001326 * pH * micro oven + 0.001104 * S/L * time − 0.00011 * S/L * micro oven + 2.31E−07 * time * micro oven − 22.2874 * pH^2 + 0.003182 * S/L^2 − 0.0025 * Time^2 − 1.3E−05 * micro oven^2.

The predicted *F*-values (3916) and *p*-values ($p < 0.0001$) show that, the model which was developed was found to be highly significant. The determination co-efficient (R^2), adjusted determination co-efficient (R^2), predicted determination co-efficient (R^2) and co-efficient of variance (CV) and signal to noise ratio (S/N)

Table 1 Box-Behnken experimental design matrix

	Factor 1	Factor 2	Factor 3	Factor 4	Response
Run	A-pH	B-S/L (ml)	C-time (s)	D-microwave power (W)	Pectin yield in (g)
1	2	20	30	540	8.67
2	2	40	60	180	8.01
3	1.5	30	60	360	14.68
4	1.5	30	45	360	15.38
5	2	40	30	540	7.64
6	1.5	30	45	360	15.2
7	1.5	20	45	360	14.67
8	1	40	30	540	11.35
9	1	40	60	540	11.95
10	1	20	60	180	7.81
11	2	20	60	540	8.37
12	1	20	30	180	7.96
13	1.5	40	45	360	16.25
14	1	20	60	540	8.14
15	2	40	30	180	7.65
16	1	30	45	360	10.57
17	1.5	30	45	360	15.19
18	1.5	30	45	540	14.78
19	1	40	30	180	11.78
20	1.5	30	45	360	15.18
21	2	20	30	180	7.88
22	1	20	30	540	8.13
23	1.5	30	45	360	15.17
24	1	40	60	180	12.45
25	2	40	60	540	7.92
26	2	30	45	360	8.57
27	1.5	30	45	360	15.17
28	2	20	60	180	7.58
29	1.5	30	30	360	14.48
30	1.5	30	45	180	14.68

was evaluated using goodness of fit model. Higher R^2 values (0.99) shows that, this model are statistically significant and this model does not explain about smaller variations (0.01). As the value of R^2 (0.99) relates with the value of *adjusted* R^2 (0.99) which shows that this model was chosen to explain the relationship between the factors and the response which was well-correlated. Lower the CV % values (0.66) states that, the deviations between experimental and predicted values are low which shows high degree of precision and reliability of the conducted experiments.

Table 2 ANOVA for response surface quadratic model of pectin yield

Source	Sum of squares	df	Mean square	*F*-value	*p*-value Prob > *F*
Model	315.5735	14	22.54097	3916.096	<0.0001
A-pH	17.70125	1	17.70125	3075.28	<0.0001
B-S/L	13.85134	1	13.85134	2406.426	<0.0001
C-time	0.104272	1	0.104272	18.11546	0.0007
D-micro oven	0.073472	1	0.073472	12.7645	0.0028
AB	17.57706	1	17.57706	3053.704	<0.0001
AC	0.074256	1	0.074256	12.90071	0.0027
AD	0.228006	1	0.228006	39.61207	<0.0001
BC	0.438906	1	0.438906	76.25223	<0.0001
BD	0.604506	1	0.604506	105.0223	<0.0001
CD	6.25E−06	1	6.25E−06	0.001086	0.9741
A^2	80.43587	1	80.43587	13974.32	<0.0001
B^2	0.262263	1	0.262263	45.56363	<0.0001
C^2	0.817863	1	0.817863	142.0893	<0.0001
D^2	0.439454	1	0.439454	76.34743	<0.0001
Residual	15.65351	15	1.043567		
Lack of fit	15.65351	10	1.565351		
Pure error	0	5	0		
Cor total	93.36667	29			

The precision (signal to noise ratio) was found to be >163.65, which denotes the best fitness of the developed models (Shanmugaprakash and Sivakumar 2013).

3.2 *Effect of pH on Pectin Yield*

In order to determine the optimized pH value for extraction, experiments were done at different pH range 1–2. From the figures (Fig. 1), it was found that the pectin yield (PY) was increasing for decreasing pH values. At lower acidic environment, solvent can easily contact with the insoluble pectin and favor the hydrolysis of insoluble pectin which results in soluble pectin, thus increasing the pectin recovery (El-Nawawi and Shehata 1988) and the maximum yield of pectin was obtained at pH 1. Also the molecular weight of pectin was reduced at low pH so that the pectin was extracted from tissues of plant (Faravash and Ashtiani 2007). However, increasing pH value beyond particular value results in reduced pectin yield (PY). The poor pectin release was due to the aggregation of pectin in the solvent. The optimum pH was 1.01, predicted by response surface methodology (Fig. 2)

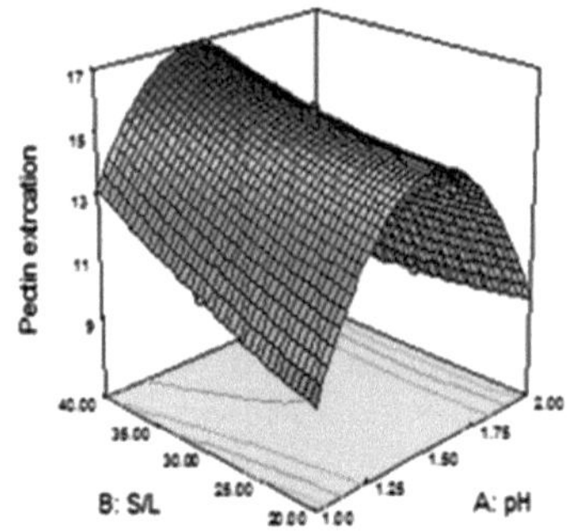

Fig. 1 Graph showing the effect of time and microwave power on the pectin yield

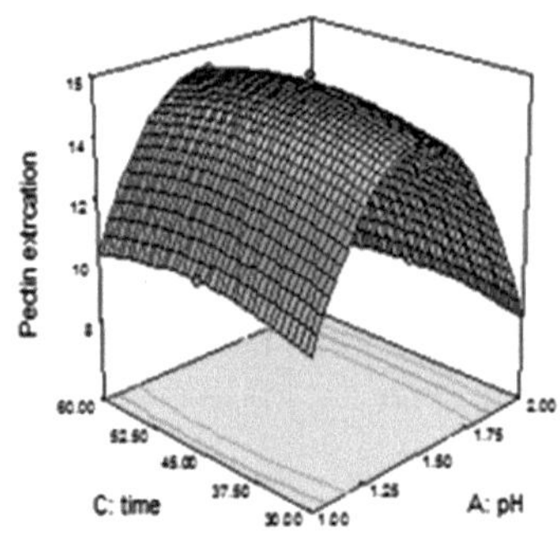

Fig. 2 Graph showing the effect of pH and time on the pectin yield

3.3 Effect of Solid-Liquid Ratio on Pectin Yield

Quantity of solvent is one of the main factors for pectin extraction. To determine the optimum extraction, sample to solvent ratio of 1:20–1:40 g/ml were used. From the results (Figs. 3, 4), the yield was increasing up to in the ratio of 1:30 g/ml. This is because of solvent can efficiently absorbs the microwave energy and plant tissues swells, which increase the contact area between plant matrix and solvent (Guo et al. 2001). So, cells were ruptured and the pectin was released. However increasing of solid-liquid ratio beyond 1:30 g/ml could decrease the extraction because of solution get saturated with higher solvent content. This could decrease the absorption of microwave energy, which affects the mass transfer rate thus by decreases the penetration and pectin yield. The optimum solid-liquid ratio was 31.06 g/ml predicted by using RSM (Fig. 3).

3.4 Effect of Time on Pectin Yield

In the extraction of pectin, time is the most important factor and used to estimate the proper emission time for the maximum pectin yield. In this work, the time range of 30–60 s investigated. From the (Fig. 4), it was found that the yield was increased at

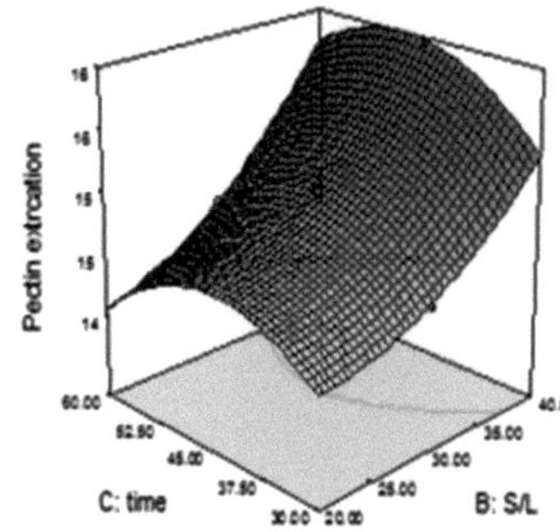

Fig. 3 Graph showing the effect of pH and microwave power on the pectin yield

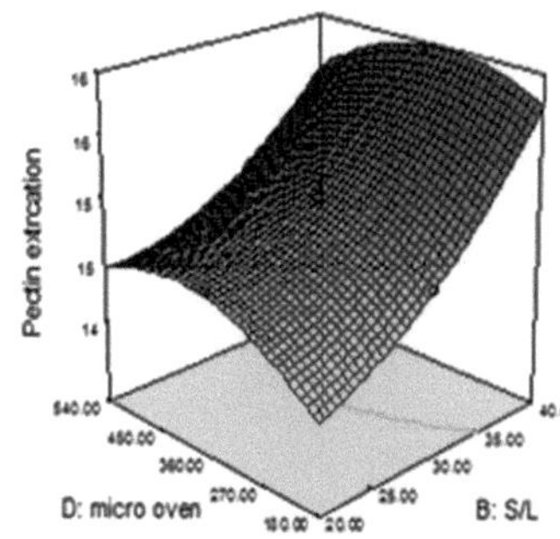

Fig. 4 Graph showing for the effect of solid-liquid ratio and time on the pectin yield

60 s. It could be observed that the dissolution of pectin into the solution occur due to the thermal accumulation of microwave energy until 60 s. The time beyond the particular limit would degrade the pectin chain. 59.7 s was the optimum time for pectin extraction predicted by RSM.

3.5 *Effect of Microwave Power on Pectin Yield*

The pectin was extracted using microwave oven. In this work different microwave powers were used in the range of 180–540 W. Here, the extraction process was carried out by heating mechanism. Microwave extraction was carried out based on two principles. One is dipole rotation and ionic conduction. Ionic conduction refers to the electrophoretic migration of the charge carriers under the influence of electric field. Dipolar rotation occurs when dipolar molecules attempts to follow the electric field. The microwave energy disrupts the cell wall of the plant material by electromagnetic radiation. So, pectin could leach out from plant material. Heat will be generated during this process. In this work higher pectin yield was obtained at 360 W and beyond the 360 W pectin could degrade due to heat generated by microwave energy. The optimum microwave power 372.49 W was predicted by RSM (Figs. 5 and 6).

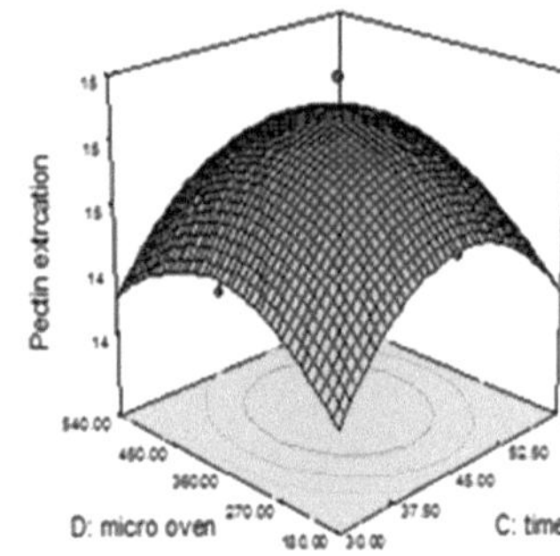

Fig. 5 Graph showing the effect of solid-liquid ratio and microwave power on the pectin yield

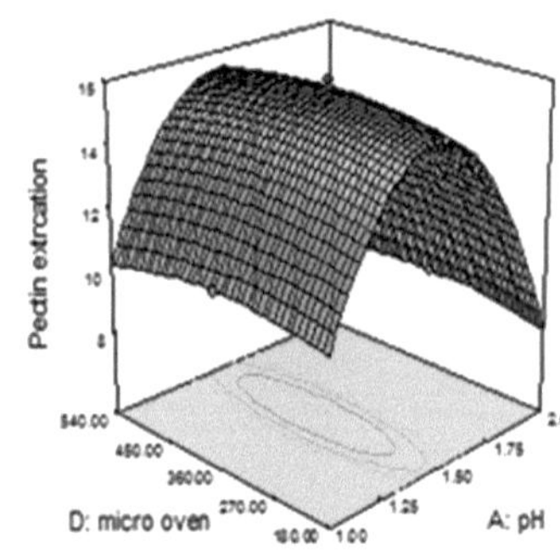

Fig. 6 Graph showing the effect of time and microwave power on the pectin yield

3.6 Determination and Validation of Optimized Conditions

To optimize pectin extraction process conditions. Derringer's desired function methodology was used. The microwave power of 482.58 W, pH of 1.37, time of 35.34 s, solid-liquid ratio of 36.15 g/100 ml gave the maximum yield of pectin was 15 % with a desirability value of 1.0. Selected optimal conditions were tested experimentally to predict the optimum response values. The experiments were performed in triplicate under the optimized conditions. The mean values (14.98 ± 0.18 %) which was obtained from the real experiments, shows the validation of the optimized conditions (15 %).

3.7 Characterization of Extracted Pectin by FT-IR

The major peak of pectin is in the range of 1000–2000 cm^{-1} (Fig. 7). The peak from 1335–1250 cm^{-1} corresponds to aromatic amines, 1650–1580 cm^{-1} corresponds to primary amines and peaks at 1650–1750 cm^{-1} corresponds to ketone stretch of the compound (Silva et al. 2012). Carboxylic acids and esters are indicated by the peak around 1350 cm^{-1} (Thetsrimuang et al. 2011). C–o and o–H

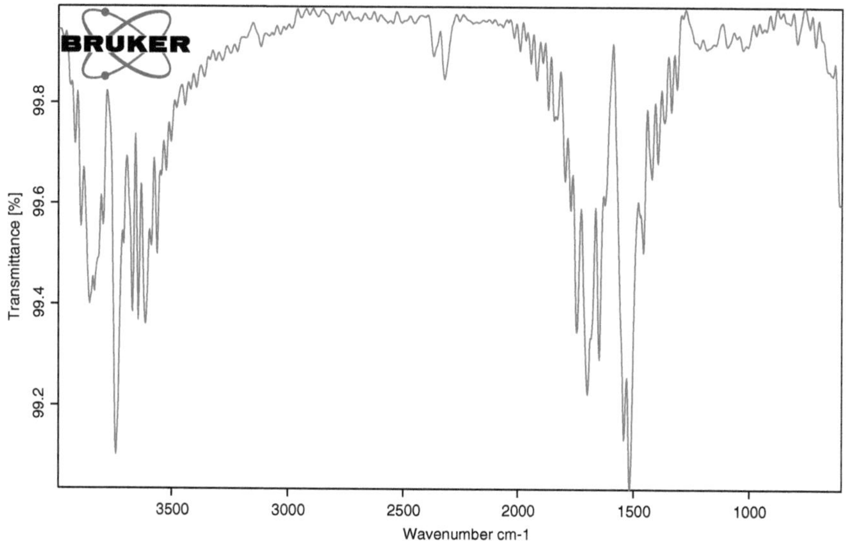

Fig. 7 FT-IR spectrum for pectin extracted from sunflower heads

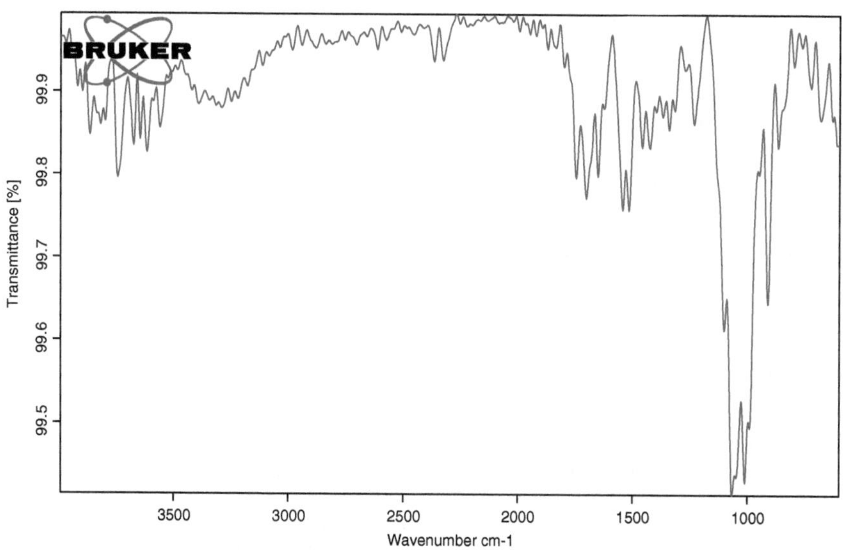

Fig. 8 FT-IR spectrum for commercial pectin

bond stretches are shown by the peaks beyond 3500 cm^{-1} (Silva et al. 2012). Moreover, the peak region between the area 1250–1728 cm^{-1} denotes the presence of α, β unsaturated esters and aliphatic amine functional group that ressembled the peak of commercial pectin (Fig. 8).

4 Conclusion

In this work, effect of microwave power, pH, time and solid-liquid ratio on the pectin yield from sunflower heads were studied. The influence of process variables increased to certain level and then decreased. The BBD model, successfully describes the inter relationship of process variables on the pectin yield. Using response surface methodology, the model was validated and the optimized conditions was found to be such as microwave power of 482.58 W, pH of 1.37, time of 35.34 s and solid-liquid ratio of 36.15 g/100 ml and the maximum yield of pectin of 15 % were obtained.

Acknowledgments The authors are thankful to the management of Kumaraguru College of Technology, Coimbatore, India, for the encouragement and providing the research facilities.

References

Agarwal, A.K., Goyal, P., Lakashminarasimhaiah, P.G., Desai, P., Singh, G.K.: Isolation of pectin from *emblica* officinalis & *Linum unitatissimum* using microwave assisted extraction technique and compare with conventional extraction method. Int. J. Pharma Res. Rev. **4**(2), 16–20 (2015)

El-Nawawi, S.A., Shehata, F.R.: Effect of the extraction temperature on the characteristics of pectin extracted from Egyptian orange peel. Biol. Wastes **24**, 307–311 (1988)

Faravash, R.S., Ashtiani, F.Z.: The effect of pH, ethanol volume and acid washing time on the yield of pectin extraction from peach pomace. Int. J. Food Sci. Technol. **42**, 1177–1187 (2007)

Fishman, M.L., Chau, H.K., Hoagland, P.D., Hotchkiss, A.T.: Microwave-assisted extraction of lime pectin. Food Hydrocolloids **20**(8), 1170–1177 (2006)

Guo, Z.K., Jin, Q.H., Fan, G.Q., Duan, Y.P., Qin, C., Wen, M.J.: Microwaveassisted extraction of effective constituents from a Chinese herbal medicine Radix puerariae. Anal. Chim. Acta **436**, 41–47 (2001)

Koh, P.C., Leong, C.M., Mohd Adzahan, N.: Microwave-assisted extraction of pectin from jackfruit rinds using different power levels. Int. Food Res. J. **21**(5), 2091–2097 (2014)

Kratchanova, M., Pavlova, E., Panchev, I.: The effect of microwave heating of fresh orange peels on the fruit tissue and quality of extracted pectin. Carbohydr. Polym. **56**(2), 181–185 (2004)

Kute, A., Mohapatra, D., Babu, B., Sawant, B.P.: Optimization of microwave assisted extraction of pectin from orange peel using response surface methodology. J. Food Res. Technol. **3**(2), 62–70 (2015)

Liu, L., Fishman, M.L., Kost, J., Hicks, K.B.: Pectin-based systems for colon-specific drug delivery via oral route. Biomaterials **24**(19), 3333–3343 (2003)

Li, J., Zu, Y.-G., Fu, Y.-J., Yang, Y.-C., Li, S.-M., Li, Z.-N., et al.: Optimization of microwave-assisted extraction of triterpene saponins from defatted residue of yellow horn (*Xanthoceras sorbifolia* Bunge.) kernel and evaluation of its antioxidant activity. Innovative Food Sci. Emerg. Technol. **11**(4), 637–643 (2010)

Maran, J.P., Manikandan, S., Thirugnanasambandham, K., VignaNivetha, C., Dinesh, R.: Box-Behnken design based statistical modeling for ultrasound-assisted extraction of corn silk polysaccharide. Carbohydr. Polym. **92**, 604–611 (2014)

Morris, G.A., Ralet, M.-C., Bonnin, E., Thibault, J.-F., Harding, S.E.: Physical characterisation of the rhamnogalacturonan and homogalacturonan fractions of sugar beet (*Beta vulgaris*) pectin. Carbohydr. Polym. **82**(4), 1161–1167 (2010)

Nwabueze, T.U.: Review article: basic steps in adapting response surface methodology as mathematical modelling for bioprocess optimisation in the food systems. Int. J. Food Sci. Technol. **45**(9), 1768–1776 (2010)

Prakash Maran, J., Mekala, V., Manikandan, S.: Modeling and optimization of ultrasound-assisted extraction of polysaccharide from *Cucurbita moschata*. Carbohydr. Polym. **92**(2), 2018–2026 (2013)

Rahmati, S., Abdullah, A., Momeny, E., Kang, O.L.: Optimization studies on microwave assisted extraction of dragon fruit (*Hylocereus polyrhizus*) peel pectin using response surface methodology. Int. Food Res. J. **22**, 233–239 (2015)

Shanmugaprakash, M., Sivakumar, V.: Development of experimental design approach and ANN-based models for determination of Cr (VI) ions uptake rate from aqueous solution onto the solid biodiesel waste residue. Bioresour. Technol. **148**, 550–559 (2013)

Shanmugaprakash, M., Kirthika, J., Ragupathy, J., Nilanee, K., Manickam, A.: Statistical based media optimization and production of naringinase using Aspergillus brasiliensis 1344. Int. J. Biol. Macromol. **64**, 443–452 (2014)

Silva, S., Martins, S., Karmali, A., Rosa, E.: Production, purification and characterisation of polysaccharides from *Pleurotus ostreatus* with antitumour activity. J. Sci. Food Agric. **92**(9), 1826–1832 (2012)

Srivastava, P., Malviya, R.: Sources of pectin, extraction and its applications in pharmaceutical industry-an overview. Indian J. Nat. Prod. Resour. **2**(1), 10–18 (2011)

Thetsrimuang, C., Khammuang, S., Chiablaem, K., Srisomsap, C., Sarnthima, R.: Antioxidant properties and cytotoxicity of crude polysaccharides from Lentinus polychrous Lév. Food Chem. **128**(3), 634–639 (2011)

Venkatesh Prabhu, M., Karthikeyan, R., Shanmugaprakash, M.: Modeling and optimization by response surface methodology and neural network–genetic algorithm for decolorization of real textile dye effluent using *Pleurotus ostreatus*: a comparison study. Desalin. Water Treat. **57**, 1–15 (2015)

Wang, S.J., Chen, F., Wu, J.H., Wang, Z.F., Liao, X.J., Hu, X.S.: Optimization of pectin extraction assisted by microwave from apple pomace using response surface methodology. J. Food Eng. **78**(2), 693–700 (2007)

Yoo, S.H., Lee, B.H., Lee, H., Lee, S., Bae, I.Y., Lee, H.G., Jr Hotchkiss, A.T., et al.: Structural characteristics of pumpkin pectin extracted by microwave heating. J. Food Sci. **77**(11), C1169–C1173 (2012)

Production and Characterization of Hydrophobins from Fungal Source

Basavaraj Hungund, Chaitanya Habib, Vaibhav Hiregoudar, Sona Umloti, Saiprasad Wandkar and Gururaj Tennalli

1 Introduction

Hydrophobins are cysteine rich proteins (about 100 amino acids) that are produced by filamentous fungi having molecular mass of around 20 kDa. These hydrophobins are reported to be produced by filamentous fungi belonging to the Ascomycetes, the Basidiomycetes and may also be produced by Zygomycetes (Scholtmeijer et al. 2001). The name hydrophobin has been used initially due to their high content of hydrophobic amino acids. Recent investigations show that, hydrophobins play crucial role in fungal growth and development. The hydrophobins are reported to reduce the surface tension of medium in which fungi grow and maintain good permeability for gaseous exchange (De Vocht et al. 1998). Hydrophobins cover the aerial spores of these filamentous fungi making their surface hydrophobic and water repellent. Based on their inter Cys spacing, hydrophobins are classified into two classes as Class I and Class II. The class I hydrophobins assemble into highly insoluble polymeric layers which are composed of fibrillar structures. These fibrillar structures are known as rodlets and possess considerable variation in the inter Cys spacing. The rodlets are extremely stable and can only be solubilized with harsh acid treatments. The soluble forms can polymerize back into rodlets under appropriate conditions. The amino acid sequence and the inter-Cys spacing are more conserved in class II hydrophobins. The monolayers formed by class II hydrophobins lack the fibrillary rodlet morphology and can be solubilized with organic solvents and detergents.

They possess a unique self-assembly mechanism and can assemble at a hydrophilic–hydrophobic interface developing a monolayer amphipathic film (Scholtmeijer et al. 2001). This property makes hydrophobins useful in many

B. Hungund (✉) · C. Habib · V. Hiregoudar · S. Umloti · S. Wandkar · G. Tennalli
Department of Biotechnology, B.V.B. College of Engineering and Technology,
Hubballi 580031, India
e-mail: hungundb@gmail.com

Prasanna B.D. et al. (eds.), *Biotechnology and Biochemical Engineering*,
DOI 10.1007/978-981-10-1920-3_5

industries such as food industry, brewing industry, enzymes immobilization, etc. Khalesi et al. (2012). The term molecular self-assembly refers to the autonomous formation of structures or patterns from pre-existing components. This is a reversible process driven only by the properties of the components (Whitesides and Grzybowski 2002). Protein self-assembly is very common in biological systems. Examples are the formation of virus capsids, the DNA replication machinery, and disease-related amyloid fibres. Hydrophobins form self-assembled structures at the air–water interface. Class I hydrophobins form a mosaic of rod-like structures, called rodlets, which are 5–10 nm in width and several hundred nanometers in length (Linder et al. 2005). Class II hydrophobins does not possess rodlet morphology. The class II hydrophobins HFBI, HFBII and HFBIII from *Trichoderma reesei* have been shown to form films with a self-assembled hexagonally ordered structure (Paananen et al. 2003). Studies on the possible interfacial self-assembly of other class II hydrophobins have not been reported. In the present investigation, an attempt was made to achieve lab scale production of hydrophobins using various fungal strains.

2 Materials and Methods

2.1 Microorganisms

For the laboratory scale production of hydrophobins, fungal strains *Trichoderma reesei* NCIM 1171, *Trichoderma viride* NCIM 1060, *Trichoderma harzianum* NCIM 1185 and an isolate-white rot fungus were used. The pure cultures of these fungal strains were maintained on Sabouraud's dextrose agar slants at 6–8 °C in the refrigerator and sub-culturing was done every three months.

2.2 Production Media

Standard medium as defined by Askolin et al. (2001) was used for the growth of fungal strains and for the production of hydrophobins. The production medium contains glucose (20 g/l), peptone (4 g/l), yeast extract (1 g/l), KH_2PO_4 (4 g/l), $(NH_4)_2SO_4$ (2.8 g/l), $MgSO_4$ (0.6 g/l), $CaCl_2{\cdot}2H_2O$ (0.8 g/l), $CoCl_2{\cdot}6H_2O$ (0.004 g/l), $MnSO_4{\cdot}H_2O$ (0.0032 g/l), $ZnSO_4{\cdot}7H_2O$ (0.0069 g/l). Initial studies were carried out using both glucose and lactose as carbon sources separately. Inoculum development was done using the standard medium and 100 ml of production medium was inoculated with each organism separately and incubated at 150 rpm at 25 °C for four days. After the completion of incubation period, the cells were harvested by centrifugation at 10,000 rpm for 15 min. The cell lysis was achieved by subjecting the cells to sonication using both bath sonicator and probe sonicator separately. Sonication process was done for 40 min to release intracellular contents along with hydrophobins. The whole contents were taken into a separating funnel and shaken vigorously to produce visible foam. The foam was verified for stability and was removed and dissolved in water. These contents were further stored at 4 °C till further testing.

2.3 *Quantitative Analysis*

The qualitative study for the hydrophobins was performed using lead sulfide test (Katoch 2011). Further quantitative analysis for hydrophobins was done by Bradford's method (1976). The amount of hydrophobins present in each sample was calculated referring to the standard graph.

2.4 *Molecular Weight Determination: SDS-PAGE*

Standard gel preparation protocol: The Resolving gel was prepared using the following, 30 % acrylamide, 10 % SDS, 10 % APS (freshly prepared), TEMED, 1.5 M Tris pH 8.8, 5× SDS Running Buffer (1 L): Tris 15 g, Glycine 72 g, SDS 5 g. Coomassie Blue Stain: 10 % (v/v) acetic acid, 0.006 %(w/v) Coomassie Blue dye, 90 % double distilled water. Stacking gel was prepared using all the same ingredients above but with 1.0 M Tris with a pH of 6.8. Then isopropanol fixing solution was prepared with 10 % (v/v) acetic acid 25 % (v/v) isopropanol 65 % double distilled water. Then 40 ml of SDS sample loading buffer was prepared using the following: double distilled water 16 ml, 0.5 M Tris pH 6.8 (5 ml), 50 % Glycerol (8 ml), 10 % SDS (8 ml), 2-mercaptoethanol (2 ml) was added immediately before using, bromophenol blue. 10 % (v/v) acetic acid.

Gel running protocol: Polyacrylamide gel was prepared according to standard protocol. The samples were loaded and gel was run at 25 mA in 1× SDS Running Buffer. Later the gel was stained with Coomassie brilliant blue and placed in a plastic container covered with isopropanol fixing solution and was kept on gel rocker at room temperature for 10 min. The fixing solution was poured off, and the gel was covered with Coomassie blue staining solution and was kept on gel rocker at room temperature for 2 h, then the staining solution was poured off and the gel was washed with 10 % acetic acid to de-stain by keeping on gel rocker at room temperature.

2.5 *Ultrafiltration*

Membrane ultrafiltration (UF) is a pressure-modified, convective process that uses semi permeable membranes to separate species in aqueous solutions by molecular size, shape, and/or charge (Schratter 2004). The different pore sizes are notated by their respective molecular weight cut offs (MWCO). In the present study MWCO of 20 KDa was used. The molecules smaller than 20 KDa pass through the membrane (permeate) whereas those larger than it remain within the membrane (retentate). The chamber was applied with positive pressure (1 bar) so that the molecules were forced through the membrane. The retentate and permeate were further analyzed for protein content by Bradford method.

3 Results and Discussion

3.1 Production of Hydrophobins

Fungal strains *Trichoderma reesei* NCIM 1171, *Trichoderma viride* NCIM 1060, *Trichoderma harzianum* NCIM 1185 and an isolate-white rot fungus were used for hydrophobin production. The cells were subjected to lysis and the contents were vigorously shaken in separating funnel. The foam produced in each case was varied for stability by exposing to air. Also the degree of foam formed and its stability was compared among all the strains. Strain *Trichoderma reesei* was found to produce the highest amount of foam with better stability (Fig. 1). The foam from this organism was dissolved into water and subjected for further qualitative and quantitative analyses.

Fig. 1 Picture showing foam produced by *Trichoderma reesei* NCIM 1171

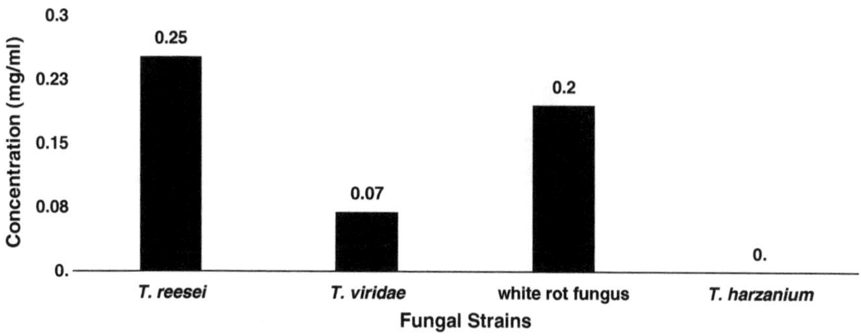

Fig. 2 Graph showing comparison of hydrophobin content from various fungal strains

The stability of the foam was measured by the time period to which the air bubbles were stable and *Trichoderma reesei* showed highest and most stable foam. More the concentration of the protein, more stable the air bubbles to process of disproportionation. The foam produced by it was taken for further analysis, Bradford estimation was performed for Quantitative analysis for all the strains and the following results were obtained. Figure 2 represents the concentrations of protein for all the fungal strains in mg/ml. The results of SDS PAGE revealed that the produced hydrophobins falls under low molecular weight proteins.

3.2 Lead Sulphide Test

The qualitative study for the hydrophobins was performed using lead sulfide test. This test is basically used to detect the presence of cysteine residue in the preparation. As hydrophobin is a cysteine rich protein the test should confirm the presence of cysteine in the given protein sample. Basically the reaction goes as follows, Cysteine rich protein upon boiling with sodium hydroxide (hot alkali), yield sodium sulphide. This reaction is due to partial conversion of the organic sulphur to inorganic sulphide, which can be detected by precipitating it to lead sulphide, using lead acetate solution.

$$\mathrm{S(Protein)} + 2\mathrm{NaOH} \longrightarrow \mathrm{Na_2S}$$
$$\mathrm{Na_2S} + (\mathrm{CH_3COO})2\mathrm{Pb} \longrightarrow \mathrm{PbS} + 2\mathrm{CH_3COONa}$$

Further quantitative analysis for hydrophobins was done by Bradford's method (1976). The amount of hydrophobins present in each sample was calculated referring to the standard graph. The foam produced was subjected to Lead sulphide test for confirmation of presence of cysteine. The formation of gray precipitate indicates positive test for presence of cysteine in the sample.

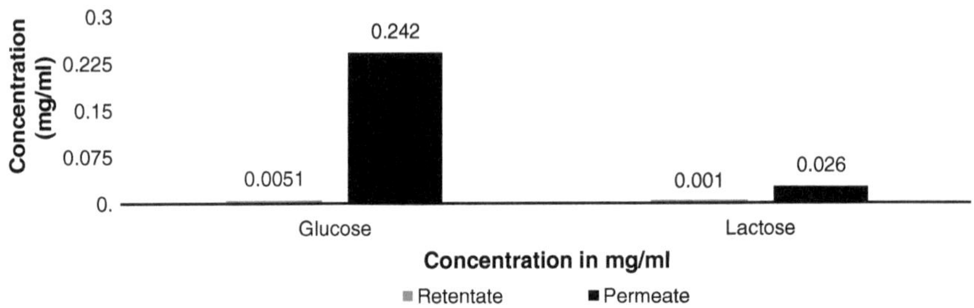

Fig. 3 Graph showing comparison of protein content in retentate and permeate for *Trichoderma reesei*

3.3 *Partial Purification by Ultrafiltration*

As the hydrophobins are smaller molecules, ultrafiltration was carried out for partial purification using 20 KDa membranes. After filtration, both permeate and retentate were taken from glucose and lactose media separately and subjected to protein estimation using Bradford method. The results show that, the protein content is more in case of permeate indicating partial purification of the protein sample. Figure 3 represent the comparison of protein content in retentate and permeate. The presence of higher amount of protein in the permeate indicates that protein may possess molecular mass less than 20 KDa. This study also demonstrates that, glucose is the preferred carbon source for the production of hydrophobins by *Trichoderma reesei* NCIM 1171.

4 Conclusion

Present investigation reports the lab scale production of fungal hydrophobins. Screening activity for hydrophobin production showed that *Trichoderma reesei* was the highest producer with stable foam. Confirmation of hydrophobins was done using cysteine residue test. Partial purification of hydrophobins from *Trichoderma reesei* was done using ultrafiltration method. Retentate and permeate obtained by ultrafiltration were subjected to Bradford assay. More concentration was observed in permeate solution which contains proteins whose molecular mass is less than 20 KDa. The results of SDS PAGE revealed that the produced hydrophobins falls under low molecular weight proteins. Thus, we have achieved lab scale production, characterization and partial purification of hydrophobins from *Trichoderma reesei*.

Acknowledgments The authors acknowledge Dr. Ashok Shettar, Vice Chancellor, KLE Technological University, Hubballi (Formerly BVBCET Hubballi) for constant support and encouragement. The authors also acknowledge Technical Education Quality Improvement Program (TEQIP) of World Bank for providing needful support to carry-out the present study.

References

Askolin, S., Nakari-Setala, T., Tenkanen, M.: Overproduction, purification, and characterization of the *Trichoderma reesei* hydrophobin HFBI. Appl. Microbiol. Biotechnol. **57**, 124–130 (2001)

Bradford, M.M.: A rapid and sensitive method for the quantitation of microgram quantities of protein utilizing the principle of protein-dye binding. Anal. Biochem. **72**, 248–254 (1976)

De Vocht, K., Van Der Scholtmeijer, E.W., De Vegte, O.M., Vries, N.Sonveaux, Wosten, A., Ruysschaert, J.M., Hadziloannou, G., Wessels, J.G., Robillard, D.G.T.: Structural characterization of the hydrophobin SC3, as a monomer and after self-assembly at hydrophobic/hydrophilic interfaces. Biophys. J. **74**, 2059–2068 (1998)

Katoch, R.: Quantitative and qualitative estimation of amino acids and proteins. Analytical Techniques in Biochemistry and Molecular Biology, pp. 93–147. Springer, Berlin (2011)

Khalesi, M., Deckers, S.M., Gebruers, K., Vissers, L., Verachtert, H., Derdelinckx, G.: Hydrophobins: exceptional proteins for many applications in brewery environment and other bio-industries. Cerevisia **37**(1), 3–9 (2012)

Linder, M.B., Szilvay, G.R., Nakari-Setala, Tiina, Penttila, Merja E.: Hydrophobins: the protein-amphiphiles of filamentous fungi. FEMS Microbiol. Rev. **29**, 877–896 (2005)

Paananen, A., Ananen, A., Vuorimaa, E., Torkkeli, M., Penttila, M., Kauranen, M., et al.: Structural hierarchy in molecular films of two class II hydrophobins. Biochemistry **42**(18), 5253–5258 (2003)

Scholtmeijer, K., Wessels, J.G.H., Wosten, H.A.B.: Fungal hydrophobins in medical and technical applications. Appl. Microbiol. Biotechnol. **56**, 1–8 (2001)

Schratter, P.: Purification and concentration by ultrafiltration. Protein Purification Protocols, vol. 244, pp. 101–116. Humana Press, New York (2004)

Whitesides, G.M., Grzybowski, B.: Self-assembly at all scales. Science **295**(5564), 2418–2421 (2002)

Sustainable Utilization of Food Industry Waste and by-Products for the Production of Prebiotic Isomaltooligosaccharides (IMO)

Anindya Basu and S.G. Prapulla

1 Introduction

Prebiotic are functional food elements that are selectively fermented by the beneficial micro-organisms in the gut and contributes to the well being of the host (Roberfroid 2007). To date, the major reported studies and the most reliable evidence accumulated for prebiotic effects have been for several non-digestible oligosaccharides (NDOs). Among these NDOs Isomaltooligosaccharides (IMO) has always received exceptional attention because of its (1) numerous physicochemical properties favourable for industrial application (Glor et al. 1988; Yoo et al. 1995; Sheu et al. 1997) and (2) several reported health benefits (Goffin et al. 2011). IMO are oligosaccharides, having mixtures of α-(1, 4) and α-(1, 6)-glycosidic linkages. Most common IMOs include panose, maltotriose, isomaltose, isomaltotriose, isomaltotetraose, kojibiose, isopanose, and other higher oligosaccharides. IMO production occurs via a complex three-step enzymatic process, whereby starch is transformed into maltose by subsequent liquefaction and saccharification. IMOs are then synthesized by using either transglucosidase or glucosyltransferase. IMOs can be used as a low calorie sweetener in foods and beverages (Goffin et al. 2011).

Food processing industries generate an enormous amount of waste causing disposal and potentially severe pollution problem. The potato processing industry produces large volume of wastes (nearly 35 % of the total potato processed). Potato processing waste (PPW) causes serious pollution problem owing to its high

A. Basu · S.G. Prapulla (✉)
Microbiology & Fermentation Technology Department, CSIR-Central Food Technological Research Institute, Mysore, India
e-mail: prapullasg@yahoo.co.in

A. Basu
e-mail: anindya.rintu@gmail.com

A. Basu · S.G. Prapulla
AcSIR—Academy of Scientific & Innovative Research, New Delhi, India

Prasanna B.D. et al. (eds.), *Biotechnology and Biochemical Engineering*,
DOI 10.1007/978-981-10-1920-3_6

chemical oxygen demand (COD) and total dissolved solid (TDS) content. In addition to protein, PPW also contains a high concentration of starch, nearly 60 % (Mironescu 2011), which can be enzymatically converted to greater value IMO reducing disposal problem. Additionally, the process can also help the farmers to tackle the hefty financial losses due to the glut of potatoes.

Food processing also leads to the generation of different by-products which are generally considered as a low-value material. One such by-product of rice milling is broken rice (BR) having 3/4 length of the whole grain. In the past, broken rice was used by beer industries. However, nowadays decent quality brewers favour whole rice in place of broken rice for brewing. Currently, broken rice is used for feeding livestock and making infant food. Alternatively broken rice can be used for IMO production as it also contains a high amount of starch (60–70 % (Gohel and Duan 2012) adding value to the otherwise considered low-value by-product of rice milling industry with better economic return.

Therefore, the current investigation aims to develop a sustainable method for the production of IMO utilizing starch-rich broken rice (BR) and potato processing waste (PPW).

2 Materials and Methods

2.1 *Chemicals and Microorganism*

IMO standards (D-maltose, maltotriose, isomaltotriose, isomaltotetraose, panose and isomaltose) were obtained from Sigma-Aldrich (USA). Acetonitrile (HPLC grade) was procured from Ranbaxy Fine Chemicals Ltd (New Delhi, India). Termamyl SC and Fungamyl 800L were a gifted by Novozyme, India. *Aspergillus niger PFS08*, a paddy field isolate was grown on potato dextrose agar slants and maintained at 4 ± 1 °C.

2.2 *Production of α-Glucosidase*

α-glucosidase was produced by submerged fermentation of *Aspergillus niger* PFS08, as described in our earlier research paper (Basu et al. 2015). The enzyme was concentrated using 50 KD Amicon filters (Sigma-Aldrich, USA). The concentrated broth, without any further purification, was used as the source of α-glucosidase in the study. The α-glucosidase assay was also performed as described by Basu et al. (2015).

2.3 Milling of BR

The rice grains were cleaned by removing foreign materials manually. The rice grains were milled using a laboratory grinder to pass through 90 mesh sieve (US standard). The rice powder was stored in an airtight plastic container at room temperature (28 ± 2 °C) until needed.

2.4 Isolation of Starch from PPW

The potato processing waste (PPW) was collected from Maxvita Foods (India) Pvt Ltd., Mysore, pH of the waste was noted, and it was stored at 4 °C until further use. The foreign materials were cleaned manually, and the solid in the waste was allowed to settle at 4 °C. The sediments were collected and washed with double distilled water several times. The sediments were then dried overnight in an oven (50 °C). The dried powder was passed through U.S standard 40 mesh sieve. The starch powder was stored in an airtight plastic container at room temperature (28 ± 2 °C) until needed.

2.5 Liquefaction

Starch (25 %) slurry was prepared, and 30 ppm Ca^{2+} (as $CaCl_2$) ions were added. The pH of the slurry was made 6.0 using 0.5 N NaOH. Termamyl SC (500 μL) was added to 500 g of starch slurry, and the mixture was maintained at 90–95 °C (in a Lab Companion water Bath) with continuous agitation (150 rpm). Samples were withdrawn at regular interval and analyzed for residual starch using KI/I_2 solution. The liquefaction process was stopped by adjusting the slurry pH to 3.0 with lactic acid.

2.6 Saccharification

The pH of the liquefied slurry was made 5.0 with the help of 0.5 N NaOH. Following the addition of 12.5 μL Fungamyl 800 L to 5 mL slurry, the reaction mixture was maintained 55 °C in a water bath (Lab Companion) with constant agitation (150 rpm) for 24 h. Sampling was done every 6 h and reaction was stopped by holding the reaction mixture in a boiling water bath for 10 min. The carbohydrate composition of the samples was found by HPLC.

2.7 Transglucosylation

The saccharified slurry (5 mL) was then adjusted to pH 4.0 with lactic acid and mixed with α-glucosidase (125 μL, 8 U/mL). The reaction mixture was held at 65 °C for 24 h in a water bath (Lab Companion) with constant agitation (150 rpm). Every 6 h, samples were collected. After arresting, the reaction mixture was analyzed by HPLC for quantification of carbohydrates.

2.8 Production of High Content IMO (HC-IMO)

The glucose and unreacted maltose in the produced IMO was removed by baker's yeast fermentation. It involved mixing of 5 % (w/v) baker's yeast with undiluted low purity oligosaccharides liquid and adjusting the pH to 5. Samples were withdrawn at a regular interval for 24 h and analysed for residual glucose and maltose by HPLC. After the yeast fermentation, resultant ethanol was removed from the liquid by evaporation.

2.9 HPLC Analysis

HPLC LC-20A (Shimadzu, Japan) installed with a refractive index detector RID 20A (Shimadzu, Kyoto, Japan) was used for the quantification of carbohydrate An aminopropyl analytical column [250 mm × 4.6 mm SS Excil amino 5 μm (Santa Clara, California)] was used with 65 % Acetonitrile as the mobile phase.

3 Results and Discussion

3.1 Liquefaction of BR and PPW

Liquefaction of both BR and PPW was complete within 4 h of reaction as indicated by the development of light red to a brown (no blue) color when reacted with KI/I_2 solution. As the liquefaction proceeded, the iodine-staining power of the slurry slowly decreased and finally almost disappeared. In liquefaction, thermostable Termamyl, SC (bacterial endo α-amylase) hydrolyses 1,4-α-glucosidic linkages in amylose and amylopectin fraction of starch producing soluble dextrins of varying chain length. Since the enzyme cannot hydrolyse the 1,6-α-glucosidic linkages in starch, the products of the action of Termamyl SC are glucose, maltose, maltotriose of panose and other higher maltooligosaccharides. The liquefied slurry from BR consisted of 22.15 ± 1.49 g/L glucose, 47.22 ± 3.18 g/L maltose, 46.17 ± 3.10 g/L maltotriose (DP3), 30.79 ± 2.07 g/L tetra-saccharides (DP4) and 43.30 ± 2.91 g/L of pentasaccharides (DP5). The HPLC analysis of the liquefied slurry produced from PPW indicated the presence of glucose (17.02 ± 1.23 g/L), maltose (41.78 ± 1.81 g/L), maltotriose (39.38 ± 2.66 g/L), DP4 (17.59 ± 0.53 g/L), DP5 (48.39 ± 3.48 g/L), and other higher oligosaccharides (36.6 ± 3.38 g/L). The results are in accordance with the other studies that showed that liquefaction of starch formed a syrup with quite a similar composition (Takasaki et al. 1991; Zanin and De Moraes 1996). It can also be noted that a high amount (40–50 %) of DP4 and other higher oligosaccharides were accumulated in the slurry which usually improve the yield of maltose during saccharification (Mironescu et al. 2009).

3.2 Saccharification of BR and PPW

Fungamyl 800L is an endo-amylase that catalyses the liberation of maltose from higher oligosaccharides. Figure 1a depicts the change in carbohydrate composition during 24 h of saccharification when the liquefied slurry from broken rice (BR) was used as the starting material. It was observed that the concentration of maltose is raised to 165.65 ± 10.09 g/L during the first 6 h of saccharification. The concentration of other sugars (DP4 and DP5) decreased gradually during this initial period of saccharification because Fungamyl 800L sequentially hydrolyzes the higher oligosaccharides to produce maltotriose (Eqs. 1–3), which is then hydrolyzed to maltose (Eq. 4). The maltotriose concentration remained constant during this time and was only found to decrease after 6 h of saccharification. The glucose concentration increased steadily throughout the saccharification and reached to 61.04 ± 2.21 g/L after 24 h of reaction. The results of saccharification also

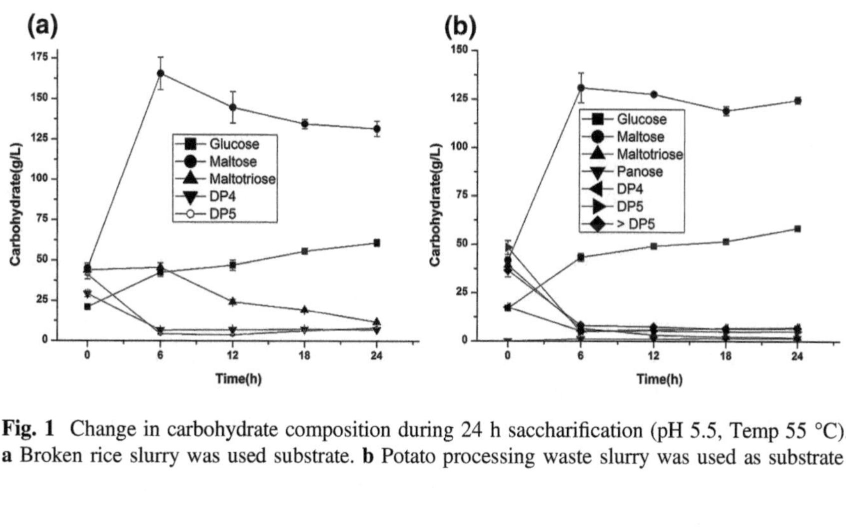

Fig. 1 Change in carbohydrate composition during 24 h saccharification (pH 5.5, Temp 55 °C). **a** Broken rice slurry was used substrate. **b** Potato processing waste slurry was used as substrate

indicated that the concentration of maltose reduced after 6 h due to its hydrolysis to glucose (Eq. 5) and hence the 6 h was selected to be optimal for saccharification of broken rice slurry.

Similar results were obtained when the liquefied slurry from potato processing waste (PPW) was used substrate for Fungamyl 800L, as shown in Fig. 1b. During initial 6 h of saccharification, the maltose concentration increased steadily to 130.92 ± 7.68 g/L whereas the concentration of other higher oligosaccharides including maltotriose declined slowly (Eqs. 1–4). The trisaccharide panose was also detected in the slurry during the saccharification that was not produced during the saccharification of broken rice slurry. The maltose concentration was maximum after 6 h of reaction and the 6 h saccharified syrup was further used for the production of IMO by transglucosylation.

The results thus obtained are compatible with the findings of Chockchaisawasdee and Poosaran (2013) who saccharified banana flour for the production of IMO.

$$G_n + H_2O \rightarrow G_{n-1} + G (n \geq 5) \quad (1)$$

$$G_5 + H_2O \rightarrow G_4 + G \quad (2)$$

$$G_4 + H_2O \rightarrow G_3^a + G \quad (3)$$

$$G_3^a + H_2O \rightarrow G_2^a + G \quad (4)$$

$$G_2^a + H_2O \rightarrow 2G \quad (5)$$

where G = Glucose, G_2^a = Maltose, G_3^a = Maltotriose, G_4 = Tetrasaccharides, G_4 = Pentasaccharides and G_4 = other higher oligosaccharides.

3.3 *Transglucosylation of BR and PPW*

Alpha-glucosidase was used to achieve transglucosylation of malto-saccharified syrup. The function of α-glucosidase is to hydrolyse α-glycosidic bonds in oligosaccharides and polysaccharides with the liberation of α-D-glucose. However, α-glucosidase also perform transglucosylation reaction at higher substrate concentration resulting in IMO (Basu et al. 2015). The IMO synthesis was observed to vary with the time of reaction and governed by Eqs. 6–9. When malto-saccharified slurry from broken rice was used as a substrate, the maximum IMO production (81.17 ± 3.15 g/L) was observed after 12 h of transglucosylation which corresponds to nearly 45 % of the total sugar (Fig. 2a and b). During this time, the concentration of glucose and panose (DP3) increased steadily while the later reached a maximum of 38.35 ± 1.64 g/L and the maltose concentration reduced considerably. During this period, glucose was released from maltose and the

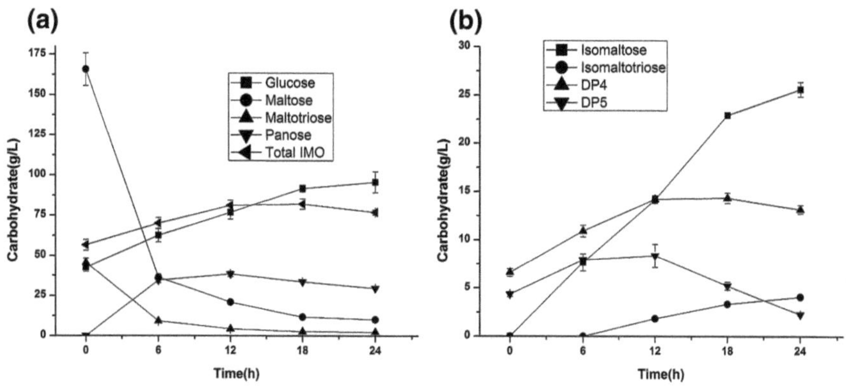

Fig. 2 **a** and **b** Time course of IMO production from malto-saccharified syrup of broken rice (pH 4, Temp 65 °C)

transfer of glucosyl residue takes place mostly to the 6-OH of the maltose and yields panose as major transglucosylation product (Eq. 6). After 6 h of reaction, the level of the tetrasaccharides (DP4) was also observed to rise due to the transfer of glucosyl residue to the 6-OH of Panose (Eq. 8). Transfer of glucosyl residue also occurred to the 6-OH of another glucose moiety leading to the formation of isomaltose (DP2) (Eq. 9) and reached 25.60 ± 0.75 g/L after 24 h of reaction. Production of isomaltotriose (DP3) was found to be initiated from 12 h of reaction as glucosyl residues are transferred to the 6-OH of the isomaltose which acts as an acceptor and increased thereafter with further incubation, reaching a maximum of 4.08 ± 0.20 g/L after 24 h of reaction (Eq. 7). The concentration of maltotriose was also found to decline whereas the concentration of DP5 increased in the course of transglucosylation.

A comparable sugar profile was obtained when the malto-saccharified slurry from PPW was used as a starting material for transglucosylation. As depicted in Fig. 3a and b, maximum IMO (89.61 ± 0.03 g/L) was produced after 18 h of transglucosylation. During this period, the main product formed was panose (31.82 ± 0.06 g/L) and isomaltose (31.36 ± 0.25 g/L). Panose was formed by transglucosylation of maltose and isomaltose was formed by transglucosylation of glucose. The production of DP4 was initiated from 12 h of reaction and reached a maximum of 13.15 ± 0.43 g/L. Isomaltotriose production did not start till 12 h and increased after that.

The results showed that transglucosylation product, in both the cases, was enriched in 1,6-α linked compounds. This could be due to their superior resistance to hydrolysis by α-glucosidase as the enzyme is more specific for the cleavage of α-(1, 4) glycosidic linkages. These observations are quite similar to that reported for the synthesis of IMO by α-glucosidase produced from *Aspergillus carbonarious* (Duan et al. 1994) and *Xantophyllomyces dendrorhous* (Fernández-Arrojo et al. 2007) which produced panose as the key transglucosylation product and *A. niger* (Duan 1995), *A. nidulans* (Kato et al. 2002) which give isomaltose as the chief

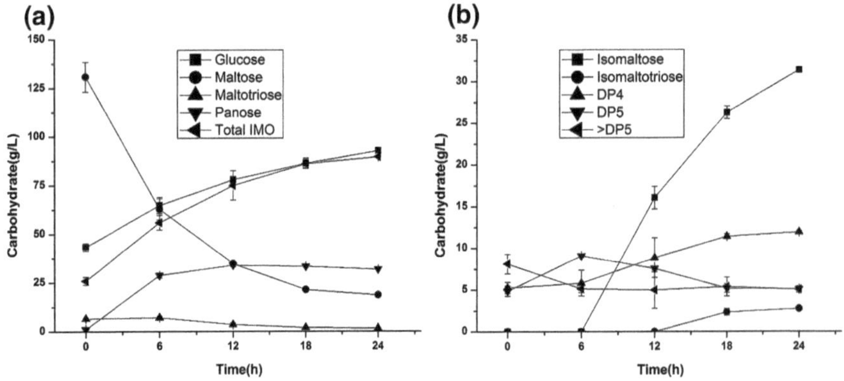

Fig. 3 **a** and **b** Time course of IMO production from the malto-saccharified syrup of potato processing waste (pH 4, Temp 65 °C)

reaction product. However, this profile contrasts the reports for α-glucosidase from *Acremonium* sp. (Konishi and Shindo 1997) that produces oligosaccharides having α-(1,2) bonds and *Paecilomyces lilacinus* (Kobayashi et al. 2003) which synthesizes both α-(1,2) and α-(1,3) bonds.

$$2G_2^a \rightarrow G_3^b + G \quad (6)$$

$$2G_2^b \rightarrow G_3^c + G \quad (7)$$

$$2G_3 \rightarrow G_4 + G \quad (8)$$

$$2G \rightarrow G_2^b \quad (9)$$

where G = Glucose, G_2^a = Maltose, G_2^b = Isomaltose, G_2^b = Panose, G_3^c = Isomaltotriose, G_4 = Tetrasaccharides.

3.4 *Production of High Content IMO*

The IMO was fermented by Baker's yeast to eliminate liberated glucose and unreacted maltose from the product Fig. 4 showed that almost all glucose and a considerable amount of maltose was removed from the oligosaccharides mixture by 12–18 h of fermentation without affecting the total IMO content. The yeast possibly failed to utilize isomaltose and other oligosaccharides due to the presence of 1,6-α-glucosidic linkages. This outcome is consistent with earlier reports that described IMO to be undigested by yeast (Pan and Lee 2005; Chockchaisawasdee and Poosaran 2013). After yeast fermentation, 92 % IMO was produced from BR and PPW composition of which is depicted in Table 1.

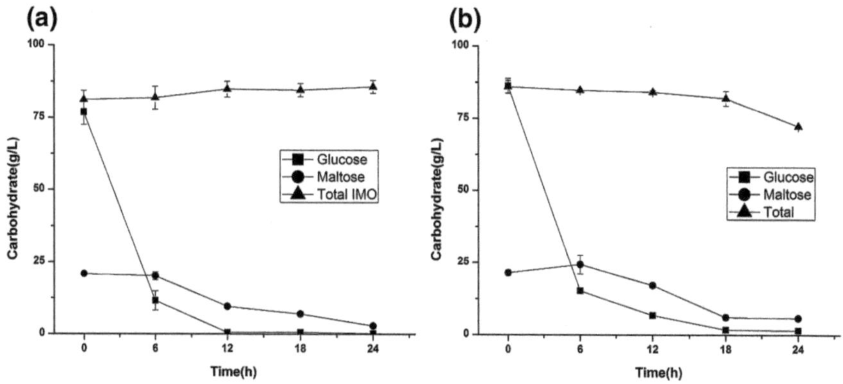

Fig. 4 Time course of Baker's yeast fermentation for the production of high content IMO, **a** from broken rice, **b** from potato processing waste

Table 1 Composition of IMO produced from BR and PPW

Component	Broken rice (BR)	Potato processing waste (PPW)
Isomaltose (g/L)	14.15 ± 0.87	26.43 ± 0.24
Maltotriose (g/L)	4.24 ± 0.18	1.89 ± 0.19
Panose (g/L)	38.13 ± 0.58	32.05 ± 2.46
Isomaltotriose (g/L)	1.98 ± 0.11	2.94 ± 0.06
DP4 (g/L)	14.57 ± 0.59	9.81 ± 0.22
DP5/Higher oligo (g/L)	11.51 ± 0.03	8.78 ± 0.19
Total IMO (g/L)	84.60 ± 2.31	81.94 ± 2.55
% IMO	92 %	92 %
Yield	0.34 kg of IMO/kg of BR	0.1 kg of IMO/l of PPW

4 Conclusion

From the present investigation, it is possible to conclude that broken rice and potato processing waste could be potential raw materials for the enzymatic synthesis of IMO. This process has the potential to solve the disposal problem of PPW to some extent. The process will also enable utilization of industrial by-products like broken rice, which are presently considered as low-value material, for the production of high-value prebiotic IMO. On the basis of the promising findings presented in this paper, investigations on the remaining issues are ongoing. We intend to concentrate on improving the yield and productivity of the process by simultaneous saccharification and transglucosylation in future. Additionally evaluating the prebiotic and physico-chemical properties of IMO mixtures is worth exploring.

Acknowledgments Anindya Basu would like to express his sincere thanks to Council of Scientific and Industrial Research (CSIR), India for the award of Fellowship. The Director, CSIR-Central Food Technological Research Institute, Mysore, India is also thanked for his support.

References

Basu, A., Mutturi, S., Prapulla, S.G.: Modeling of enzymatic production of isomaltooligosaccharides: a mechanistic approach. Catal. Sci. Technol. **5**, 2945–2958 (2015)

Chockchaisawasdee, S., Poosaran, N.: Production of isomaltooligosaccharides from banana flour. J. Sci. Food Agric. **93**, 180–186 (2013)

Duan, K.: Transglucosylation of a fungal α-glucosidase. Ann. N. Y. Acad. Sci. **750**, 325–328 (1995)

Duan, K.J., Sheu, D.C., Lin, M.T., Hsueh, H.C.: Reaction-mechanism of isomaltooligosaccharides synthesis by alpha-glucosidase from *Aspergillus carbonarious*. Biotechnol. Lett. **16**, 1151–1156 (1994)

Fernández-Arrojo, L., Marín, D., Gómez De Segura, A., et al.: Transformation of maltose into prebiotic isomaltooligosaccharides by a novel α-glucosidase from *Xantophyllomyces dendrorhous*. Process Biochem. **42**, 1530–1536 (2007)

Glor, E.B., Miller, C.H., Spandau, D.F.: Degradation of starch and its hydrolytic products by oral bacteria. J. Dent. Res. **67**, 75–81 (1988)

Goffin, D., Delzenne, N., Blecker, C., et al.: Will isomalto-oligosaccharides, a well-established functional food in Asia, break through the European and American market? The status of knowledge on these prebiotics. Crit. Rev. Food Sci. Nutr. **51**, 394–409 (2011)

Gohel, V., Duan, G.: No-cook process for ethanol production using Indian broken rice and pearl millet. Int. J. Microbiol. 1–9 (2012)

Kato, N., Suyama, S., Shirokane, M., et al.: Novel alpha-glucosidase from *Aspergillus nidulans* with strong transglycosylation activity. Appl. Environ. Microbiol. **68**, 1250–1256 (2002)

Kobayashi, I.K., Okuda, M.T., Ashimoto, H.H., et al.: Purification and characterization of a new type of α-glucosidase from *Paecilomyces lilacinus* that has transglucosylation activity to produce α-1, 3- and α- 1, 2-linked oligosaccharides. Biosci. Biotechnol. Biochem. **67**, 29–35 (2003)

Konishi, Y., Shindo, K.: Production of nigerose, nigerosyl glucose, and nigerosyl maltose by *Acremonium* sp. S4G13. Biosci. Biotechnol. Biochem. **61**, 439–442 (1997)

Mironescu, M.: Investigations on wastewaters at potato processing and starch recovery and characterisation. J. Agroalim. Process Technol Investig. **17**, 134–138 (2011)

Mironescu, M., Mironescu, I.D., Trifan, A.: Influence of the liquefied starch composition and pH on the saccharification at the obtaining of maltose syrup. Bull. UASVM Agric. **66**, 364–369 (2009)

Pan, Y.C., Lee, W.C.: Production of high-purity isomalto-oligosaccharides syrup by the enzymatic conversion of transglucosidase and fermentation of yeast cells. Biotechnol. Bioeng. **89**, 797–804 (2005)

Roberfroid, M.: Prebiotics: the concept revisited. J. Nutr. **137**, 830S–837S (2007)

Sheu, D.C., Huang, C.I., Duan, K.J.: Production of isomaltooligosaccharides by alpha-glucosidase immobilized in chitosan beads and by polyethyleneimine-glutaraldehyde treated mycelia of *Aspergillus carbonarious*. Biotechnol. Tech. **11**, 287–291 (1997)

Takasaki, Y., Shinohara, H., Tsuruhisa, M., Imada, K.: Maltotetraose-producing amylase from *Bacillus*. Agric. Biol. Chem. **55**, 1715–1720 (1991)

Yoo, S.H., Kweon, M.R., Kim, M.J., et al.: Branched oligosaccharides concentrated by yeast fermentation and effectiveness as a low sweetness humectant. J. Food Sci. **60**, 516–519 (1995)

Zanin, G.M., De Moraes, F.F.: Modeling cassava starch saccharification with amyloglucosidase. Appl. Biochem. Biotechnol. **57**(58), 617–625 (1996)

Enhancing the Bioproduction of Cellulase by *Aspergillus nidulan* via Medium Optimization

P. Saravanan and R. Muthuvelayudham

1 Introduction

Enzymatic hydrolysis of cellulose to glucose was carried out by the enzyme cellulase, a multi enzyme complex made up of several proteins. The cellulase enzyme is synthesized by microorganisms for their growth on cellulosic materials. The microorganisms are aerobic, anaerobic, mesophilic (or) thermophilic. Among them, the genera of *Clostridium*, *Cellulomonas*, Thermomonospora, *Trichoderma*, *Aspergillus* are extensively studied cellulase producer (Riswanali et al. 2012). The fungus *Aspergillus nidulan* is an efficient producer of cellulase enzymes.

The lignocelluloses are agro industrial products by solid materials and utilized as physical support and source of nutrients in Solid state fermentation (SmF). The lignocellulosic substrates are such as barley bran, oat straw, sugarcane bagasse, grape, rice straw, corn, wheat straw, and others (Singhania et al. 2009; Dashtban et al. 2009). These biodegradable wastes are recognized as potential sustainable source for production of various value added products like, biofuel, animal feed, chemicals and enzymes (Ping et al. 2011). Fruit wastes fabricated in large amounts in India are both used as an animal feed and disposed of to the soil. Since fruit wastes include large amounts of carbohydrates and some nutrients, they can function as substrates for production of ligninolytic enzymes and other value-added products due to their renewable nature and low-cost availability through SmF (Akpinar and Ozturk Urek 2012).

P. Saravanan (✉) · R. Muthuvelayudham
Department of Chemical Engineering, Annamalai University, Annamalai Nagar, Chidambaram 608002, Tamil Nadu, India
e-mail: pancha_saravanan@yahoo.com

R. Muthuvelayudham
e-mail: muthupriya_03@yahoo.com

Prasanna B.D. et al. (eds.), *Biotechnology and Biochemical Engineering*,
DOI 10.1007/978-981-10-1920-3_7

The present study was aimed at screening of nutrients and optimization of the selected nutrients in SmF using Plackett–Burman method and central composite design (CCD) to enhance the cellulase production.

2 Materials and Methods

2.1 Microorganisms and Maintenance

A. nidulan were obtained from the Institute of Microbial Technology, Chandigarh, India. The cultures were maintained on potato dextrose agar slants at 4 °C and the slants were sub-cultured by every month.

2.2 Inoculum Preparation

The inoculum preparation, 2.0 ml of a spore suspension (containing 10^8 conidia/ml) of *A. nidulan* was inoculated into 50 ml of seed medium in 250 ml Erlenmeyer flask and cultured at 30 °C, pH of 5.5 and 180 rpm for three days.

2.3 Fermentation Conditions

Fermentation was carried out in 250 ml cotton plugged Erlenmeyer flasks with 10 g of pre-treated pomegranate peel (Saravanan et al. 2012) at pH 7. The supplemented with different nutrient concentration for tests according to the selected factorial design and sterilized at 120 °C for 20 min. After cooling the flasks at room temperature, the flasks were inoculated with 1 ml of grown culture broth. The flasks were maintained at 30 °C under agitation at 200 rpm for 48 h. During the preliminary screening process, the experiments were carried out for 9 days and it was found that the maximum production was obtained at 6th day. Hence further experiments were carried out for 6 days.

2.4 Enzyme Assay

Cellulase activity (measured as filter paper hydrolysing activity, using a 1 × 6 cm strip of Whatman no. 1 filter paper) and cellobiase activity were assayed by Ghose (Ghose 1987) and expressed as international units (IU). One international unit of cellulase activity is the amount of enzyme that forms 1 μmol glucose (reducing sugars as glucose) per minute during the hydrolysis reaction. Reducing sugar was determined by the dinitro salicylic acid (DNS) method (Ghose 1987).

2.5 *Optimization of Cellulase Production*

Design of experiment (DOE)

The RSM combined with a 3^3 full factorial experimental design was used to point out the relationship existing between the response functions and effect of nutrients as well as to determine the conditions of these variables able to optimize the fermentation.

3 Results and Discussions

Initially the carbon and nitrogen sources were screened and among them the best carbon and nitrogen sources were selected for further optimization using PB design. PB design was adopted to optimize various medium components for the production of cellulase fermentation by *A. nidulan*. Various media components were investigated for their effect in the process of cellulase production. The Pareto plots offer a convenient view of the results obtained by PB design. The main effects plot is very useful in determining the cellulase production at intermediate levels of different combination of the independent variable. From the Pareto chart (Fig. 1) the

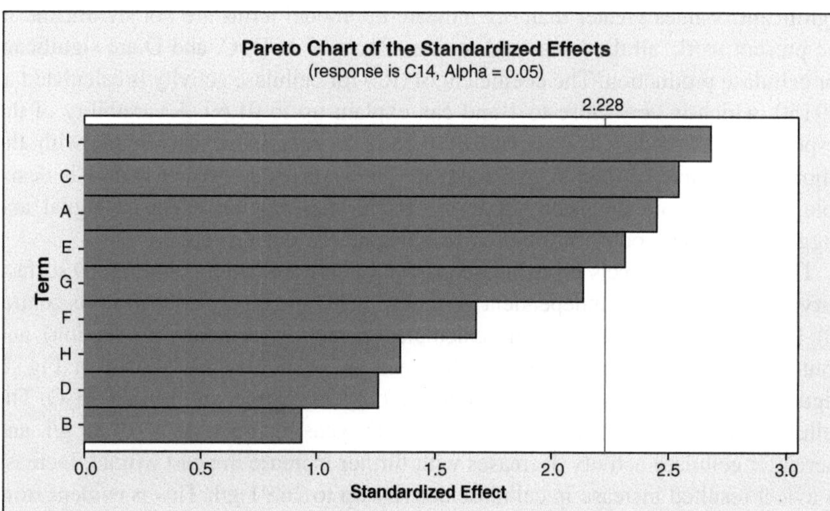

Fig. 1 Pareto chart (*A*—avicel, *B*—corn steep flour, *C*—$MnSO_4{\cdot}H_2O$, *D*—$FeSO_4{\cdot}7H_2O$, *E*—beef extract, *F*—soybean cake flour, *G*—KH_2PO_4, *H*—$CoCl_2{\cdot}6H_2O$, *I*—yeast extract)

Table 1 Ranges of variables used in RSM

Variables (g/l)	Code	−2	−1	0	+1	+2
Yeast extract	I	5	7.5	10	12.5	15
$MnSO_4{\cdot}H_2O$	C	0.6	0.8	1.0	1.2	1.4
Avicel	A	15	20	25	30	35
Beef extract	E	0.4	0.6	0.8	1.0	1.2

variables viz., yeast extract, $MnSO_4{\cdot}H_2O$, Avicel and beef extract, were selected for further optimization to attain a maximum response (Table 1).

The level of factors yeast extract, $MnSO_4{\cdot}H_2O$, Avicel and beef extract and the effect of interactions on cellulase production were determined by central composite design of RSM. Thirty experiments were preferred at different proportions of the factors shown in (Table 2) and the central point was repeated five times (8, 17, 20, 21 and 26). The results were analysed by ANOVA. The second order regression equation provided the levels of cellulase activity.

$$\begin{aligned} \mathrm{Y} = {} & 11.63 + 0.43\mathrm{I} + 0.94\mathrm{C} + 0.53\mathrm{A} + 1.05\mathrm{E} - 0.67\mathrm{IC} \\ & + 0.20\mathrm{IA} + 0.13\mathrm{IE} + 0.86\mathrm{CA} + 0.087\mathrm{BE} - 0.24\mathrm{CE} \\ & - 0.60I^2 - 0.95C^2 - 1.27A^2 - 0.91E^2 \end{aligned} \tag{1}$$

where Y is the cellulase yield (IU/mL), I, C, A and Y is yeast extract, $MnSO_4{\cdot}H_2O$, Avicel and beef extract respectively.

The analysis of variance of the quadratic regression method demonstrated for the highly significant model. The Model F-value of 11.69 implies that the model is significant. There is only a 0.01 % chance that a "Model F-Value" this large could occur due to noise. Values of "Prob > F" less than 0.05 indicate model terms are significant. Values greater than 0.1 indicate the model terms are not significant. In the present work, all the linear and square effects of A, B, C, and D are significant for cellulase production. The coefficient of (R^2) for cellulase activity is calculated as 0.9160, which is very close to 1 and can explain up to 91.60 % variability of the response. The predicted R^2 value of 0.5882 is reasonable agreement with the adjusted R^2 value of 0.8376. An adequate precision value greater than 4 is desirable. The adequate precision value of 10.079 indicates an adequate signal and suggests that the model can be used to navigate the design space.

The interactive effects for cellulase production are studied by plotting 3D surface curves against any two independent variables, while the other variables at its central (0) level. The 3D curves of the calculated response (cellulase production) and contour plots from the interactions between the variables are shown in Fig. 2. Figure 2 shows the dependency of cellulase on yeast extract and $MnSO_4{\cdot}H_2O$. The cellulase activity increases with increase in yeast extract upto 10.35 g/l and thereafter cellulase activity decreases with further increase in yeast extract. Increase in avicel resulted increase in cellulase activity up to 26.91 g/l. This is evident from Fig. 2 show the dependency of cellulase activity on beef extract. The effect of beef extract on cellulase observed is similar to other variables. The maximum cellulase

Table 2 Central composite design (CCD) of factors in coded levels with enzyme activity

Run no.	I	C	A	E	Experimental cellulase activity	Predicted cellulase activity
1	−1	1	−1	1	7.9	5.23
2	0	0	0	2	11.6	6.50
3	−1	−1	−1	1	7.1	4.28
4	1	−1	−1	−1	6.3	5.23
5	0	0	0	−2	5.1	4.17
6	0	−2	0	0	6.8	4.48
7	1	1	−1	1	8.5	5.49
8	0	0	0	0	12.3	7.80
9	1	−1	1	−1	7.8	4.89
10	0	0	0	0	11.3	7.80
11	0	2	0	0	9.6	6.38
12	1	1	1	1	10.7	7.71
13	1	−1	−1	1	8.9	5.64
14	0	0	2	0	7.5	6.70
15	−1	1	1	1	11.5	7.69
16	−1	1	1	−1	9.6	5.78
17	0	0	0	0	12.3	7.80
18	−1	−1	1	−1	4.6	3.73
19	−2	0	0	0	9.4	4.47
20	0	0	0	0	12.2	7.80
21	0	0	0	0	11.2	7.80
22	1	−1	1	1	8.8	6.46
23	−1	−1	−1	−1	5.3	3.81
24	2	0	0	0	9.8	5.90
25	1	1	−1	−1	6.3	4.78
26	0	0	0	0	10.5	7.80
27	0	0	−2	0	6.3	4.57
28	1	1	1	−1	9.4	5.84
29	−1	−1	1	1	4.6	5.34
30	−1	1	−1	−1	6.5	4.46

activity is observed at 0.90 g/l of beef extract. Figure 2 shows the dependency of cellulase activity on $MnSO_4{\cdot}H_2O$.

The maximum cellulase activity was observed at 1.12 g/l afterwards, the organism exhibited a decreasing trend. The optimum conditions for the maximum production of cellulase are estimated by regression equation. The optimum conditions are: yeast extract—10.35 g/l, $MnSO_4{\cdot}H_2O$—1.12 g/l, Avicel—26.91 g/l and beef extract—0.90 g/l.

The interactive effects for cellulase production are studied by plotting 3D surface curves against any two independent variables, while the other variables at its central (0) level. The 3D curves of the calculated response (cellulase production) and

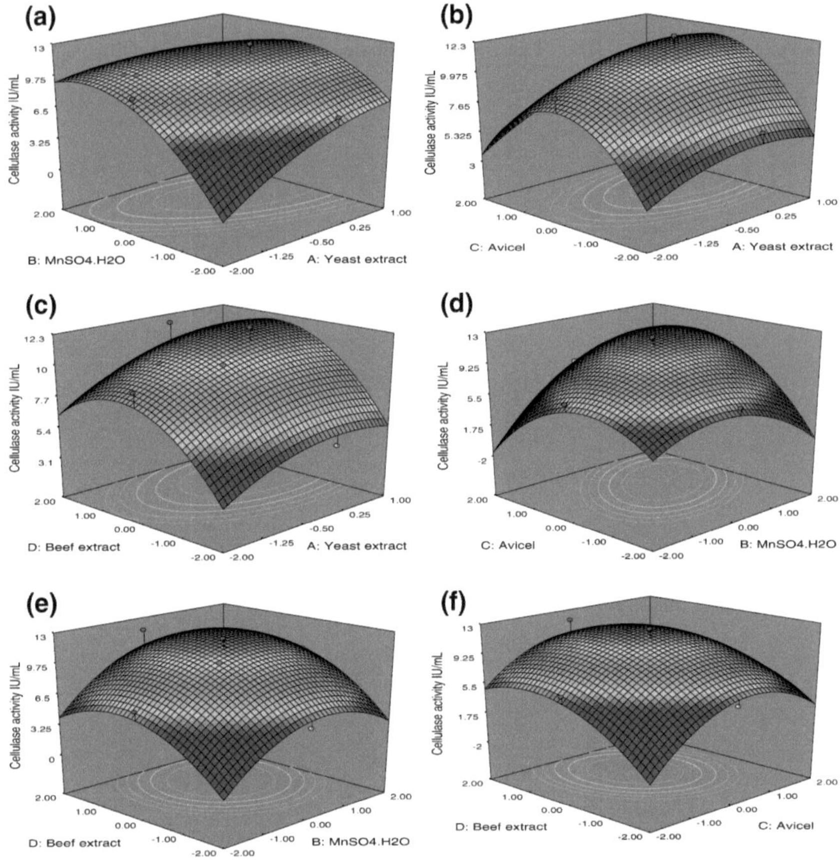

Fig. 2 *3D* plot showing the effect of nutrients on the cellulase production

contour plots from the interactions between the variables are shown in Fig. 2a–f. Figure 2a shows the dependency of cellulase on yeast extract and $MnSO_4{\cdot}H_2O$. The cellulase activity increases with increase in yeast extract upto 10.35 g/l and thereafter cellulase activity decreases with further increase in yeast extract. The similar observation in Fig. 2b. Increase in avicel resulted increase in cellulase activity up to 26.91 g/l. This is evident from Fig. 2c, e and f show the dependency of cellulase activity on beef extract. The effect of beef extract on cellulase observed is similar to other variables. The maximum cellulase activity is observed at 0.90 g/l of beef extract. Figure 2d shows the dependency of cellulase activity on $MnSO_4{\cdot}H_2O$. The maximum cellulase activity was observed at 1.12 g/l afterwards, the organism exhibited a decreasing trend. The optimum conditions for the maximum production of cellulase are estimated by regression equation. The optimum conditions are: yeast extract—10.35 g/l, $MnSO_4{\cdot}H_2O$—1.12 g/l, Avicel—26.91 g/l and beef extract—0.90 g/l.

4 Conclusions

The production of an enzyme by a bioprocess using a fruit waste is a value addition to the fruit residues. In this work, optimization of cellulase production by *Aspergillus nidulan* using the cheap substrate pomegranate waste was tried. The nutrients were screened using PBD and the selected nutrients were optimized using CCD to enhance the cellulase production. The best conditions for enzyme production are identified as yeast extract—10.35 g/l, $MnSO_4{\cdot}H_2O$—1.12 g/l, Avicel—26.91 g/l, and beef extract—0.90 g/l with respect to media components. Using the optimized conditions, the cellulase production reaches 11.6 IU/mL. Additionally, these findings are potential for developing an eco-friendly process for industries, involving in the processing of lignocellulosic substrates containing wastes to confer value added products.

References

Akpinar, M., OzturkUrek, R.: Production of ligninolytic enzymes by solid state fermentation using *Pleurotus eryngii*. Prep. Biochem. Biotechnol. **42**, 582–597 (2012)

Dashtban, M., Schraft, H., Qin, W.: Fungal bioconversion of lignocellulosic residues; opportunities and perspectives. Int. J. Biol. Sci. **5**(6), 578–595 (2009)

Ghose, T.K.: Measurement of cellulase activities. Pure Appl. Chem. **59**(2), 257–268 (1987)

Ping, L., Brosse, N., Sannigrahi, P., Ragauskas, A.: Evaluation of grape stalks as a bioresource. Ind. Crops Prod. **33**, 200–204 (2011)

Riswanali, S.B., Saravanan, P., Muthuvelayudham, R., Viruthagiri, T.: Optimization of nutrient medium for cellulase and hemicellulase productions from rice straw: a statistical approach. Int. J. Chem. Anal. Sci. **3**(4), 1364–1370 (2012)

Saravanan, P., Muthuvelayudham, R., Viruthagiri, T.: Application of statistical design for the production of cellulase by *Trichoderma reesei* using mango peel. Enzyme Res. Article ID 157643, 7 pages (2012)

Singhania, R.R., Patel, A.K., Zoclo, C.R., Pandey, A.: Recent advances in solid-state fermentation. Biochem. Eng. J. **44**, 13–18 (2009)

Modeling and Theoretical Analysis of Isomaltooligosaccharide (IMO) Production Using Fed-Batch Process

Sarma Mutturi and Anindya Basu

1 Introduction

A prebiotic is defined as a 'non-digestible food ingredient that beneficially affects the host by selectively stimulating the growth and/or activity of one or a limited number of bacteria in the colon and thus improves host health' (Gibson and Roberfroid 1995). Non-digestible oligosaccharides (NDOs) are drawing an increasing amount of attention as prebiotics since health promoting effects of some of this prebiotics are well documented. Oligosaccharides with a degree of polymerization (DP) between 2 and 8 have shown to selectively increase the growth of *lactobacillus* sp. and *bifidobacteria* sp. Oligosaccharides are also helpful as they are not used by pathogenic microbes such as *Salmonella* and *Escherichia coli* in the colon (Licht et al. 2012). Prebiotics are mainly associated with breakfast cereals, baked goods, cereal bars and baby foods, as well as some dairy products. The probiotics market has come into existence primarily on the launch of special yogurt and fermented milk drinks and is growing at a fast rate.

Isomaltooligosaccharides (IMO) is a promising prebiotic widely used for food or feed additives because of their mild sweetness superior heat and acid stability, lower viscosity and water activity (Glor et al. 1988; Yoo et al. 1995; Sheu et al. 1997). Structurally IMO is oligosaccharides (DP 2-5) containing glucose monomer linked together by α-(1,4) and α-(1,6) glyosidic linkages. The most abundant IMOs are isomaltose, panose, isomaltotriose, isomaltotetraose, isopanose and higher branched oligosaccharides. Although traditionally glucosyl-transferase (EC 2.4.1) is used for the conversion of maltose to isomaltooligosaccharides, the widely avail-

S. Mutturi (✉) · A. Basu
Microbiology and Fermentation Technology Department,
CSIR-CFTRI, Mysore 570020, India
e-mail: sarma.mutturi@gmail.com

S. Mutturi · A. Basu
AcSIR—Academy of Scientific and Innovative Research, New Delhi, India

Prasanna B.D. et al. (eds.), *Biotechnology and Biochemical Engineering*,
DOI 10.1007/978-981-10-1920-3_8

able α-glucosidase (EC 3.2.1.20) can also be used. It hydrolyses α-glycosidic linkages from the non-reducing end of oligosaccharides and polysaccharides with the release of α-glucose. Besides this hydrolytic activity, the enzyme also displays a transferase activity at higher maltose concentration (transglucosylation reaction), which results in the formation of IMO (Pan and Lee 2005).

In our previous study, we described a batch process for the production of IMO by enzymatic transglucosylation of maltose using α-glucosidase obtained from *Aspergillus niger* PFS (Basu et al. 2015). The reaction mechanism was formulated and modeled mechanistically using a set of experimental data and later validated against a different experimental setup in the same study. The results showed several inhibition parameters were significant and lowered the yield of the process considerably during transglucosylation. Interestingly in a fed-batch system such inhibitions can be avoided by adding the substrate (maltose) gradually to the reactor. Therefore in the present study our earlier comprehensive model for the batch system has been extended to a fed-batch system and theoretical analysis have been carried out. The model simulations were carried for two different feeding strategies, i.e., constant feeding and linear feeding and some simulations have indeed resulted in higher yields in comparison to the batch experiments. To the best of our knowledge, this is the first study to deal with IMO production in a fed batch system.

2 Methods

2.1 Model Development

IMO produced from maltose involves a complex combination of simultaneous hydrolysis and transglucosylation reactions in an enzymatic setup. The model was developed using inventory equation wherein the dynamic accumulation rates are accounted as follows:

Rate of accumulation of the state component = Input of the component (via feeding) + generation of the component (via hydrolysis/transglucosylation)—consumption of the component (via hydrolysis/transglucosylation)

The rate of accumulations of individual components is as follows:

Glucose:

$$\frac{dG}{dt} = \left(\frac{2\times 180}{342}\right)(r_1 + r_4) + \left(\frac{180}{504}\right)(r_2 + r_3 + r_5 + r_{14}) + \left(\frac{180}{666}\right)(r_6 + r_7) + \left(\frac{180}{2\times 342}\right)(r_8 + r_9 + r_{10}) - r_{13} - \left(\frac{F(t)}{V(t)}\right)G$$

Maltose:

$$\frac{dG_2}{dt} = -r_1 + \left(\frac{342}{504}\right)(r_3 + r_5) - (r_8 + r_9) + \left(\frac{342}{2 \times 504}\right)(r_{11} + r_{12}) + \left(\frac{F(t)}{V(t)}\right)(S_F - G_2)$$

Isomaltose:

$$\frac{dG_2'}{dt} = \left(\frac{342}{504}\right)(r_2 + r_{14}) - r_4 - r_{10} + \left(\frac{342}{2 \times 180}\right)r_{13} - \left(\frac{F(t)}{V(t)}\right)(G_2')$$

Panose:

$$\frac{dG_3}{dt} = -(r_2 + r_3) + \left(\frac{504}{666}\right)r_7 + \left(\frac{504}{2 \times 342}\right)r_9 - r_{12} - \left(\frac{F(t)}{V(t)}\right)(G_3)$$

Maltotriose:

$$\frac{dG_3'}{dt} = -r_5 + \left(\frac{504}{666}\right)r_6 + \left(\frac{504}{2 \times 342}\right)r_8 - r_{11} - \left(\frac{F(t)}{V(t)}\right)(G_3')$$

Isomaltotriose:

$$\frac{dG_3''}{dt} = -r_{14} + \left(\frac{504}{2 \times 342}\right)r_{10} - \left(\frac{F(t)}{V(t)}\right)(G_3'')$$

Tetraose:

$$\frac{dG_4}{dt} = -r_6 + \left(\frac{666}{2 \times 504}\right)r_{11} - \left(\frac{F(t)}{V(t)}\right)(G_4)$$

Glucosylpanose:

$$\frac{dG_4'}{dt} = -r_7 + \left(\frac{666}{2 \times 504}\right)r_{12} - \left(\frac{F(t)}{V(t)}\right)(G_4')$$

Flowrate:

$$F = \frac{dV}{dt}$$

In case of constant flowrate $F(t)$ = constant, where as in case of linear flowrate $F(t)$ = slope × time (t)
The reaction terms are defined in Table 1.

Table 1 Kinetics assumed for reaction rates during modeling of fed-batch process

$r_1 = \frac{v_{1r} \times G_2}{K_{1M} + \left(1 + \frac{G_2}{K_{1iG_2}}\right) G_2}$	$r_2 = \frac{v_{2r} \times G_3}{K_{2M} + G_3}$	$r_3 = \frac{v_{3r} \times G_3}{K_{3M}\left(1 + \frac{G}{K_{3iG}}\right)\left(1 + \frac{G_2}{K_{3iG_2}}\right) + G_3}$
$r_4 = \frac{v_{4r} \times G_2'}{K_{4M}\left(1 + \frac{G}{K_{4iG}}\right) + G_2'}$	$r_6 = \frac{v_{6r} \times G_4}{K_{6M}\left(1 + \frac{G}{K_{6iG}}\right) + G_4}$	$r_5 = \frac{v_{5r} \times G_3'}{K_{5M}\left(1 + \frac{G}{K_{5iG}}\right)\left(1 + \frac{G_2}{K_{5iG_2}}\right) + G_3'}$
$r_7 = \frac{v_{7r} \times G_4'}{K_{7M}\left(1 + \frac{G}{K_{7iG}}\right) + G_4'}$	$r_8 = \frac{v_{8r} \times G_2}{K_{8M} + G_2\left(1 + \frac{G_2}{K_{8iG}}\right)}$	$r_9 = \frac{v_{9r} \times G_2}{K_{9M} + G_2\left(1 + \frac{G_2}{K_{9iG}}\right)}$
$r_{10} = \frac{v_{10r} \times G_2'}{K_{10M}\left(1 + \frac{G}{K_{10iG}}\right) + G_2'}$	$r_{11} = \frac{v_{11r} \times G_3'}{K_{11M}\left(1 + \frac{G_2}{K_{11iG_2}}\right) + G_3'}$	$r_{12} = \frac{v_{12r} \times G_3}{K_{12M}\left(1 + \frac{G_2}{K_{12iG_2}}\right) + G_3}$
$r_{13} = \frac{v_{13r} \times G}{K_{13M} + G}$	$r_{15} = F(t)$	

The parameters are defined in the appendix section. The parameter values were obtained from Basu et al. (2015) which were computed using genetic algorithm while calibrating against the experimental data.

2.2 *Simulations*

All the simulations were carried using MATLAB® 2013b on Windows 8.1pro platform with Intel(R) Core(TM) i7-4770 processor. The solution to a set of ODEs was obtained using the MATLAB ODE solver, ode15s. This solver uses variable order and variable step-length with implicit numerical differentiation formula (NDF) to compute the solution over each time interval (Ashino et al. 2000).

3 Results and Discussion

3.1 *Constant Feeding*

The fed-batch simulations for IMO production using constant flow rate were carried using different values as shown in Table 2. The comparison in final yield of IMO with reference to the batch reaction and also in terms of other variables is shown in Table 2. It can be observed that above the feed rate values of 0.005 L/h the yields in IMO is higher than of batch reaction. Moreover from the values it can be observed

Table 2 Comparison of batch and constant feeding fed-batch production of IMO

Time (h)	Mode	Flow rate (L/h)	Initial volume (L)	Final volume (L)	Total IMO (g/L)	Final glucose (g/L)	Residual maltose (g/L)	Yield predicted (g IMO/g maltose)
24 h	Batch	0	–	–	319.57	265.58	23.23	0.55
	Constant feeding	0.0005	0.01	0.022	166.33	107.62	55.85	0.31
		0.005	0.01	0.13	289.64	166.87	100.70	0.58
		0.05	0.01	1.21	312.07	175.39	111.24	0.64
		0.5	0.01	12.01	314.47	176.17	112.53	0.65
60 h	Batch	0	–	–	227.32	380.28	8.69	0.38
	Constant feeding	0.0005	0.01	0.04	207.68	214.41	35.28	0.37
		0.005	0.01	0.31	280.22	263.51	45.74	0.51
		0.05	0.01	3.01	290.48	269.46	47.17	0.53
		0.5	0.01	30.01	291.56	270.06	47.32	0.53

Fig. 1 Yield at different constant flow rates for enzymatic reaction time of 24 and 60 h

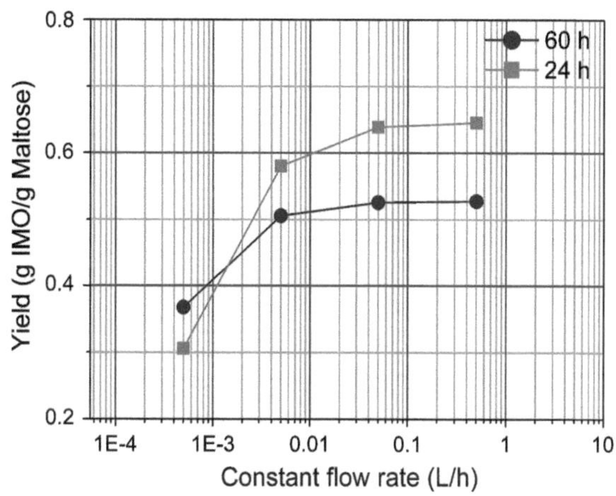

that the yields levelled beyond the constant flow rate of 0.05 L/h (Fig. 1). Thus, it can be concluded that with operating flow rates between 0.005 and 0.5 L/h would alleviate substrate inhibition during IMO production. The time profiles of all the components at 0.005 L/h constant feed rate is shown in Fig. 2. Here the volume follows a linear profile for a constant feed rate profile.

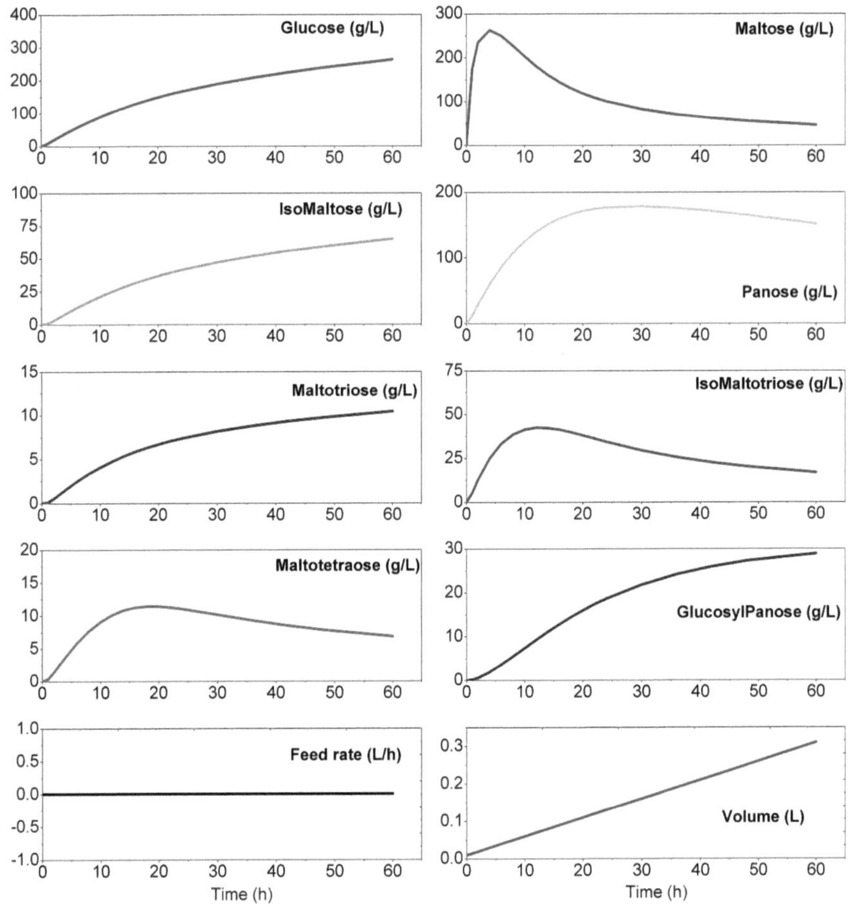

Fig. 2 Time profiles of state variables during constant feed (0.005 L/h) fed-batch of IMO production

3.2 *Linear Feeding*

Here in linear feeding, the feed profile follows a linear curve with a constant slope along the time axis. The slopes were predetermined and the time course profiles for the state variables have been observed at each selected slope. The yield of total IMO during different linear feeding profiles has been computed and shown in Table 3. It can be observed from this table that the yields of IMO could reach a value of 0.68, higher than that of batch and constant feed rate fed-batch. Also it can be observed from Table 3 that as the slope increased beyond 2E−03 L/h^2, the yields were consolidated to 0.66–0.68 and 0.58–0.59 in case of 24 and 60 h, respectively. It can be observed that the substrate inhibition can be effectively alleviated using linear feed fed-batch production IMO. Figure 3 describes the effect of slope of

Table 3 IMO response values for linear feeding fed-batch production of IMO

Time (h)	Start flow rate (L/h)	Final flow rate (L/h)	Slope	Initial volume (L)	Final volume (L)	Total IMO (g/L)	Final glucose (g/L)	Residual maltose (g/L)	Yield predicted (g IMO/g Maltose)
Linear Feeding (24 h)	0.005	0.0055	2E−05	0.01	0.016	103.80	59.33	57.46	0.19
	0.005	0.0098	2E−04	0.01	0.068	241.40	122.15	149.37	0.54
	0.005	0.053	2E−03	0.01	0.59	274.08	133.38	184.24	0.66
	0.005	0.485	2E−02	0.01	5.78	277.38	134.03	189.63	0.68
	0.005	4.805	2E−01	0.01	57.70	277.70	134.06	190.24	0.68
Linear feeding (60 h)	0.005	0.0062	2E−05	0.01	0.046	230.00	182.92	61.66	0.43
	0.005	0.017	2E−04	0.01	0.37	294.89	216.79	78.02	0.56
	0.005	0.1205	2E−03	0.01	3.62	303.69	220.62	80.23	0.58
	0.005	1.205	2E−02	0.01	36.08	304.63	221.00	80.47	0.59
	0.005	12.005	2E−01	0.01	360.75	304.72	221.04	80.49	0.59

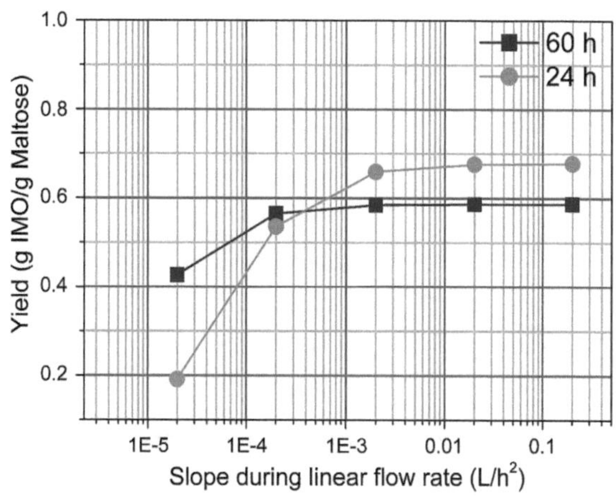

Fig. 3 Yield at different constant flow rates for enzymatic reaction time of 24 and 60 h

linear feed on IMO yield. It can be observed with increasing slope in the range of 2E−5 to 0.002 L/h^2 the yield of IMO has increased for both reaction times of 24 and 60 h. However beyond 0.002 L/h^2 the yield of IMO has not increased significantly and reached a plateau with maximum yields of 0.68 and 0.59 for 24 and 60 h of reaction times, respectively (Fig. 3).

The time course analysis of one of the slopes (2E−03 L/h^2) is provided in Fig. 4. From this figure it can be observed that the volume follows a parabolic profile for a linear feed profile.

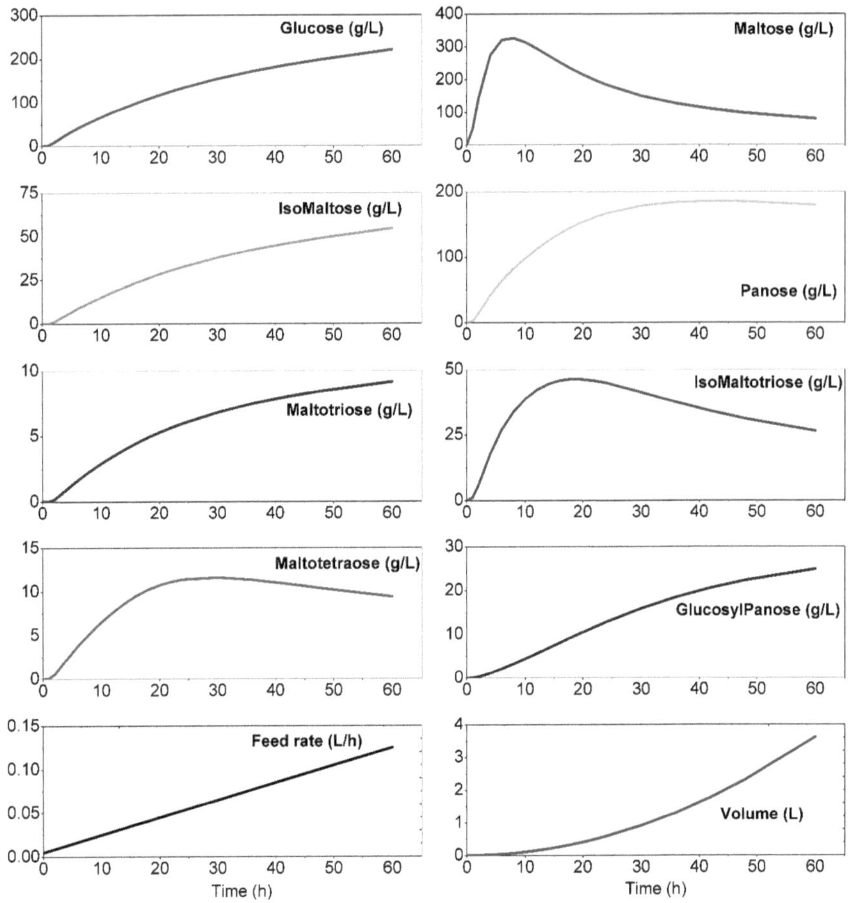

Fig. 4 Time profiles of state variables during linear feed (2E−03 L/h^2) fed-batch of IMO production

4 Conclusion

Using 0.005 and 0.05 L/h constant feeding for 60 and 24 h fed-batch reactions, respectively, would be feasible in a lab scale reactor to achieve optimal IMO yield with final working volume reaching 0.31 and 1.21 L, respectively. Similarly for linear feeding the 2×10^{-4} and 2×10^{-3} L/h^2 slopes would be appropriate for bench scale reactors, where the final volumes would reach 0.37 and 0.59 L, respectively. Both linear and constant feed-based fed-batch production of IMO has alleviated substrate and product inhibitions significantly and thus contributed increased yield and productivity of IMO when compared to batch studies.

Acknowledgments The authors would like to thank Director, CSIR-CFTRI and Head, Microbiology and Fermentation Technology, CSIR-CFTRI for their kind support and encouragement.

Appendix

G	Glucose	V_{4r}	Maximum velocity of hydrolysis of isomaltose to glucose (g/L h)
G_2	Maltose	$\mathbf{K_{4M}}$	Michaelis-Menton constant of hydrolysis of isomaltose to glucose (g/L)
G_2'	Isomaltose	$\mathbf{K_{4iG}}$	Competitive inhibition constant by glucose on isomaltose as substrate (g/L)
G_3	Panose	v_{5r}	Maximum velocity of hydrolysis of maltotriose to maltose (g/L h)
G_3'	Maltotriose	$\mathbf{K_{5M}}$	Michaelis-Menton constant of hydrolysis of maltotriose to maltose (g/L)
G_3''	Isomaltotriose	$\mathbf{K_{5iG}}$	Competitive inhibition constant by glucose on maltotriose as substrate (g/L)
G_4	Maltotetraose	$\mathbf{K_{5iG2}}$	Competitive inhibition constant bymaltose on maltotriose as substrate (g/L)
G_4'	Glucosyl-panose	v_{6r}	Maximum velocity of hydrolysis of maltotetraose to maltotriose (g/L h)
v	Maximum velocity (g/L h)	$\mathbf{K_{6M}}$	Michaelis-Menton constant of hydrolysis of maltotetraose to maltotriose (g/L)
$\mathbf{K_M}$	Michaelis-Menton constant (g/L)	$\mathbf{K_{6iG}}$	Competitive inhibition constant by glucose on maltotetraose as substrate (g/L)
$\mathbf{K_i}$	Inhibition constant (g/L)	v_{7r}	Maximum velocity of hydrolysis of glucosyl-panose to panose (g/L h)
R	Rate of reaction (1, 2, 3, … subscripts denotes Eqs. 1, 2, 3, …)	$\mathbf{K_{7M}}$	Michaelis-Menton constant of hydrolysis of glucosyl-panose to panose (g/L)
v_{1r}	Maximum velocity of hydrolysis of maltose to glucose (g/L h)	$\mathbf{K_{7iG}}$	Competitive inhibition constant by glucose on glucosyl-panose as substrate (g/L)
$\mathbf{K_{1M}}$	Michaelis-Menton constant of hydrolysis of maltose to glucose (g/L)	v_{8r}	Maximum velocity of hydrolysis of isomaltotriose to isomaltose (g/L h)

(continued)

(continued)

K_{1iG2}	Substrate inhibition constant of hydrolysis of maltose to glucose (g/L)	K_{8M}	Michaelis-Menton constant of hydrolysis of isomaltotriose to isomaltose (g/L)
v_{2r}	Maximum velocity of hydrolysis of panose to isomaltose (g/L h)	v_{9r}	Maximum velocity of transglucosylation of maltose to maltotriose (g/L h)
K_{2M}	Michaelis-Menton constant of hydrolysis of panose to isomaltose (g/L)	K_{9M}	Michaelis-Menton constant of transglucosylation of maltose to maltotriose (g/L)
v_{3r}	Maximum velocity of hydrolysis of panose to maltose (g/L h)	K_{9iG2}	Substrate inhibition constant of transglucosylation of maltose to maltotriose (g/L)
K_{3M}	Michaelis-Menton constant of hydrolysis of panose to maltose (g/L)	v_{10r}	Maximum velocity of transglucosylation of maltose to panose (g/L h)
K_{3iG}	Competitive inhibition constant by glucose on panose as substrate (g/L)	K_{10M}	Michaelis-Menton constant of transglucosylation of maltose to panose (g/L)
K_{3iG2}	Competitive inhibition constant by maltose on panose as substrate (g/L)	K_{10iG2}	Substrate inhibition constant of transglucosylation of maltose to panose (g/L)
v_{11r}	Maximum velocity of transglucosylation of isomaltose to isomaltotriose (g/L h)	K_{13M}	Michaelis-Menton constant of transglucosylation of panose to glucosyl-panose (g/L)
K_{11M}	Michaelis-Menton constant of transglucosylation of isomaltose to isomaltotriose (g/L)	K_{13iG2}	Competitive inhibition constant by maltose on panose as substrate (g/L)
K_{11iG}	Substrate inhibition constant of transglucosylation of isomaltose to isomaltotriose (g/L)	v_{14r}	maximum velocity of transglucosylation of glucose to isomaltose (g/L h)
v_{12r}	Maximum velocity of transglucosylation of maltotriose to maltotetraose (g/L h)	K_{14M}	Michaelis-Menton constant of transglucosylation of glucose to isomaltose (g/L)
K_{12M}	Michaelis-Menton constant of transglucosylation of maltotriose to maltotetraose (g/L)		
K_{12iG2}	Substrate inhibition constant of transglucosylation of maltotriose to maltotetraose (g/L)		
v_{13r}	Maximum velocity of transglucosylation of panose to glucosyl-panose (g/L h)		

References

Ashino, R., Nagase, M., Vaillancourt, R.: Behind and beyond the MATLAB ODE Suite. Comput. Math. Appl. **40**(4–5), 491–512 (2000)

Basu, A., Mutturi, S., Prapulla, S.G.: Modeling of enzymatic production of isomaltooligosaccharides: a mechanistic approach. Catal. Sci. Technol. **5**, 2945–2958 (2015)

Gibson, G.R., Roberfroid, M.B.: Dietary modulation of the human colonic microbiota: introducing the concept of prebiotics. J. Nutr. **125**, 1401–1412 (1995)

Glor, E.B., Miller, C.H., Spandau, D.F.: Degradation of starch and its hydrolytic products by oral bacteria. J. Dent. Res. **67**, 75–81 (1988)

Licht, T.R., Ebersbach, T., Frøkiær, H.: Prebiotics for prevention of gut infections. Trends Food Sci. Technol. **23**, 70–82 (2012)

Pan, Y.C., Lee, W.C.: Production of high-purity isomalto-oligosaccharides syrup by the enzymatic conversion of transglucosidase and fermentation of yeast cells. Biotechnol. Bioeng. **89**, 797–804 (2005)

Sheu, D.C., Huang, C.I., Duan, K.J.: Production of isomaltooligosaccharides by alpha-glucosidase immobilized in chitosan beads and by polyethyleneimine-glutaraldehyde treated mycelia of Aspergillus carbonarious. Biotechnol. Tech. **11**, 287–291 (1997)

Yoo, S.H., Kweon, M.R., Kim, M.J., Auh, J.H., Jung, D.S., Kim, J.R., Yook, C., Kim, J.W., Park, K.H.: Branched oligosaccharides concentrated by yeast fermentation and effectiveness as a low sweetness humectant. J. Food Sci. **60**, 516–519 (1995)

Optimization of Na_2CO_3 Pre-treatment by RSM Approach for Releasing Reducing Sugars from Cocoa Pod Shells

Vinayaka B. Shet, Nisha, Manasa Bhat, Manasa, Leah Natasha S. Mascarenhas, Louella C. Goveas, C. Vaman Rao and P. Ujwal

1 Introduction

Lignocelullose is the most abundant energy source on earth (Wang et al. 2007). Biological conversion of lignocellulose into biochemicals involves hydrolysis of polysaccharides such as cellulose and hemicelluloses into monosaccharides via enzyme or chemicals prior to fermentation (Badger 2002). Lignin prevents cellulase to catalyse cellulose and also adsorbs enzymes, making them inactive for hydrolysis (Converse et al. 1990). Alkali pre-treatments are among the most effective methods due to their ability to remove lignin (Zhang et al. 2012). Pre-treatment at mild conditions using inexpensive chemicals is suggested to decrease the cost of pre-treatment (Karimi et al. 2013). Sodium carbonate is one of the inexpensive and environmentally friendly alkali chemicals used for pre-treatment, under harsh conditions and also oxidative pre-treatment of lignocellulosic materials (Bjerre et al. 1996; Salehi et al. 2012). Cocoa pod was reported to contain about 43.9–45.2 % carbohydrate (Davendra 1980). Cocoa pod shell are abundantly available in Dakshina Kannada (D.K) India. In the current study one factor at a time (OFAT) analysis was carried out. Further parameter levels identified by OFAT were used to generate centre points for response surface methodology (RSM) experiments to achieve maximum production of reducing sugar as response.

V.B. Shet (✉) · Nisha · M. Bhat · Manasa · L.N.S. Mascarenhas ·
L.C. Goveas · C.V. Rao · P. Ujwal
Department of Biotechnology Engineering, NMAMIT (V.T.U, Belagavi),
Nitte 574110, Karnataka, India
e-mail: vinayakabshet@nitte.edu.in

Prasanna B.D. et al. (eds.), *Biotechnology and Biochemical Engineering*,
DOI 10.1007/978-981-10-1920-3_9

2 Methodology

2.1 Processing of Raw Material

The raw material used in the study was Cocoa pod shells (CPS) which were sundried and then further moisture content was removed by drying in hot air oven at around 100 °C. Size reduction unit operation was carried out to achieve powder of CPS by using grinder. All chemicals, solvents and reagents used in the study were of analytical grade, unless mentioned.

2.2 Optimization of CPS Pre-treatment

2.2.1 Selection of Significant Parameters and Their Levels by OFAT

Pre-treatment of CPS was carried out by sodium carbonate hydrolysis. The OFAT approach was used to select the significant physical parameters and the levels for pre-treatment process are given in Table 1.

Estimation of reducing sugars released was carried out by dinitrosalicylic acid (DNSA) method (Miller 1972) after neutralising the samples. The parameter at which maximum reducing sugars released were chosen as the centre point values to enhance the pre-treatment process by RSM.

2.2.2 Central Composite Design

The factors (X1 and X2) exhibiting significant effect during OFAT studies were chosen and their levels were optimized for maximum extraction of reducing sugars as response (Y1) using central composite design (CCD) for Na_2CO_3 hydrolysis. CCD was designed for RSM optimization by STATISTICA software (Statsoft 1999) with two factors and five levels (Table 2) and it consisted of 12 experimental runs (Table 3).

The data of pre-treatment studies obtained was subjected to analysis by using analysis of variance (ANOVA) using the same software.

The optimization experiments were carried out as per Tables 3. The levels of other non significant factors, agitation speed and weight of CPS was maintained at

Table 1 Parameters and their levels for OFAT studies

Parameter	Notation	Test range
Weight of CPS (%w/v)	X1	1–7
Na_2CO_3 concentration (N)	X2	1–6
Time (h)	X3	1–6
Agitation speed (rpm)	X4	50–150

Table 2 Coded parameters and their actual levels for optimisation by CCD

Factors	Notations	Levels				
		−α	−1	0	1	+α
Weight (%w/v)	X1	2.171	3	5	7	7.828
Concentration (N)	X2	0.585	1	2	3	3.414

Table 3 CCD for optimization of CPS pre-treatment

Run	X1	X2	Y1
1	3	1	0.985
2	3	3	0.631
3	7	1	0.884
4	7	3	0.530
5	2.17	2	0.783
6	7.82	2	0.934
7	5	0.58	1.237
8	5	3.41	0.631
9(C)	5	2	0.682
10(C)	5	2	0.682
11(C)	5	2	0.732
12(C)	5	2	0.682

100 rpm and 0.5 g/L (from OFAT studies). The expected products of hydrolysis are glucose, galactose, cellobiose and minor saccharides as xylose and arabinose. All of these saccharides are reducing sugars (Ranjan et al. 2014).

2.3 *Validation of the Second Order Polynomial Model*

The second order polynomial model obtained from RSM was validated. The experimental output was then compared to the values predicted by the second order model obtained from CCD, to find out the goodness of fit of the model.

3 Results

3.1 *Optimisation of Pre-treatment of CPS*

3.1.1 Selection of Significant Parameters and Their Levels by OFAT

The significant physical parameters and the initial test range of four factors (Table 1) for the pre-treatment processes were obtained by OFAT. When X1 was

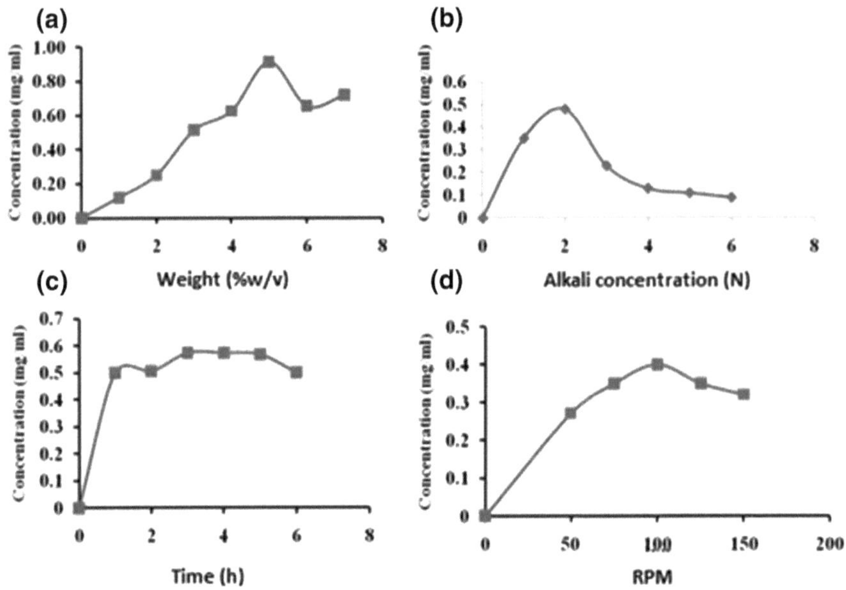

Fig. 1 Effect of change in **a** CPS weight; **b** Na_2CO_3 concentration (N); **c** time (h) and **d** agitation speed (rpm) on the release of reducing sugars by Na_2CO_3 hydrolysis of CPS

varied by keeping other factors constant (X_2 = 2 N, X_3 = 1 h, X_4 = 100 rpm), maximum released reducing sugar (RRS) concentration of X1 was 5 (%w/v) represented in Fig. 1a. Similarly X2 was 2 N (Fig. 1b). In the present study, on variation of time and agitation speed, it was observed that, there was no significant difference in RRS (Fig. 1c, d). On the basis of these results, levels of factors (X_1 and X_2) were determined and subjected to optimization by CCD.

3.1.2 Optimisation of Parameters for Release of Reducing Sugars by Alkali Hydrolysis by CCD

The influence of X1 and X2 on release of reducing sugars was determined by CCD as indicated in Table 3.

Table 4 shows that only the linear effect of the Na_2CO_3 showed highly significant effect on the reducing sugar concentration released by alkali hydrolysis of CPS (Y1) whereas all the other effects were found to be insignificant.

It was observed that RRS decreased with increase in X2 and increased with increase in X1 (Fig. 2). On the basis of desirability plots the optimized conditions for obtaining maximum RRS (Y1) were 7.82 g of CPS treated with 0.58 N of Na_2CO_3 (Fig. 3). Concentration of RRS was found to be 0.94 g/L at optimized condition.

Table 4 ANOVA table showing the effect of independent and interaction effects of variables on the RRS by Na_2CO_3 hydrolysis of CPS

	SS	df	MS	F	P
X1(L)	1.71E−05	1	1.71E−05	0.00154	0.96993
X1(Q)	0.01433	1	0.01433	1.32776	0.29304
X2(L)	**0.30616**	**1**	**0.30616**	**28.3749**	**0.00178**
X2(Q)	0.04631	1	0.04631	4.29185	0.08369
X1 * X2	0	1	0	0	1
Error	0.06474	6	0.01079		
Total SS	0.42334	11			

Values less than 0.05 indicate significance at 95% confidence interval and are emphasized in bold

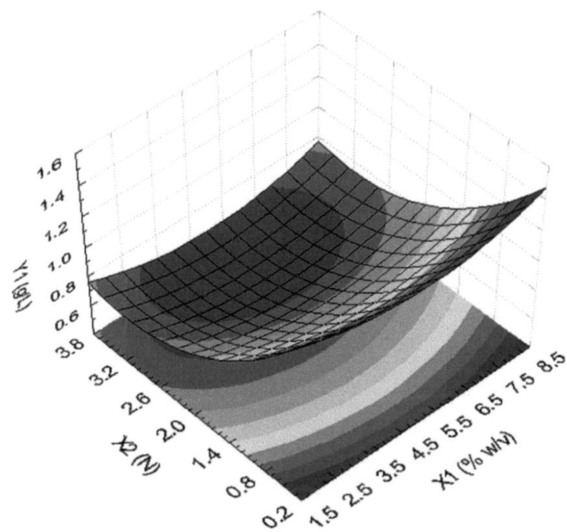

Fig. 2 3D plot showing the variation of RRS with respect to changes in X1 and X2

The regression equation for RRS by alkali hydrolysis, as a function of the two independent variables (X_1 and X_2) and their linear and quadratic interactions, are represented by the following model

$$Y1 = 1.718 - 0.1179X_1 - 1.536X_2 + 0.0118X_1^2 + 0.0851X_2^2 + 2.49E - 17X_1X_2$$

Coefficient of determination (R^2) was found to be 0.84, is a measure of the strength of the linear relationship between the experimental and predicted value.

The observed values of RRS were compared with the RRS values as predicted by the second order models. These results of Table 5 indicates that there is excellent correlation between experimental and predicted values and in turn proves the validity of the models.

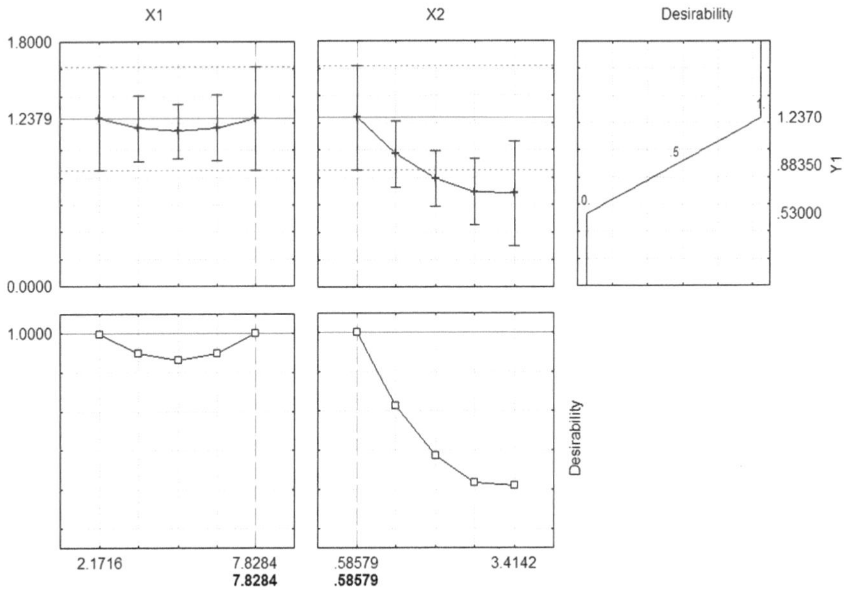

Fig. 3 Profiles for desirability levels of factors X1 and X2 for maximum RRS by Na_2CO_3 hydrolysis

Table 5 Validation experiments indicating the fitness of the model generated by CCD to predict the RRS from pre-treated CPS

RUN #	X1	X2	Yex	Ypr
1	2.5	1.5	0.71	0.885
2	6	2.5	0.56	0.63
3	4	3	0.51	0.595
4	5	3	0.51	0.583
5	7	2	0.71	0.743
6	3	1	0.81	1.021

4 Conclusion

Optimization studies in the field of pre-treatment of lignocellulosic biomass are scarce. In this context, an attempt was made to optimize the release of reducing sugars by Na_2CO_3 hydrolysis of cocoa pod shells by using CCD. RSM was found to be efficient in describing the effect of significant factors and modeling the pre-treatment process. The significance of this study is that the optimized conditions obtained from this study can be used for the release of reducing sugars from the lignocellulosic raw material which could be used as an input for production of biochemical's by fermentation.

Acknowledgments Authors thank Mr. Anantharamakrishna from the Peruvai village of Vittla (D.K.) India for providing cocoa pod shells as per the requirement and VGST, Government of Karnataka, India for funding the project.

References

Badger, P.C.: Ethanol from cellulose: a general review. In: Janick, J., Whipkey, A. (eds.) Trends in New Crops and New Uses. ASHS Press, Alexander (2002)

Bjerre, A.B., Olesen, A.B., Fernqvist, T., Ploger, A., Schmidt, A.S.: Pre-treatment of wheat straw using combined wet oxidation and alkaline hydrolysis resulting in convertible cellulose and hemicelluloses. Biotechnol. Bioeng. **49**, 568–577 (1996)

Converse, A.O., Ooshima, H., Burns, D.S.: Kinetics of enzymatic hydrolysis of lignocellulosic materialsbased on surface area of cellulose accessible to enzyme and enzyme adsorption on lignin and cellulose. Appl. Biochem. Biotechnol. **24–25**(1), 67–73 (1990)

Davendra, C.: The feeding value of by-product from cocoa and coconuts in diets for farm livestock. In: Proceeding of the International Conference on Cocoa and Coconuts, Kuala Lumpur, 457–471 (1980)

Karimi, K., Shafiei, M., Kumar, R.: Progress on physical and chemical pre-treatment of lignocellulosic biomass. In: Kumar Gupta, V. (ed.), Biofuels and biorefineries: recent developments. Springer Science Publisher, Germany, pp. 53–96 (2013)

Miller, G.L.: Use of dinitrosalicylic acid reagent for determination of reducing sugar. Anal. Chem. **31**(3), 426–428 (1972)

Rajan, K., Carrier, D.J.: Effect of dilute acid pre-treatment conditions and washing on the production of inhibitors and on recovery of sugars during wheat straw enzymatic hydrolysis. Biomass Bioenergy **62**, 222–227 (2014)

Salehi, S.M.A., Karimi, K., Behzad, T., Poornejad, N.: Efficient conversion of rice straw to bioethanol using sodium carbonate pre-treatment. Energy Fuels **26**, 7354–7361 (2012)

Statsoft: Statistics for Windows, Statsoft Inc., Tulsa, USA (1999)

Wang, M., Wu M, Huo H.: Life cycle energy and greenhouse gas emission impacts of different corn ethanol plant types. Environ. Res. Lett. **2**,1–13 (2007)

Zhang, Sh, Keshwani, D.R., Xu, Y., Hanna, M.A.: Alkali combined extrusion pre-treatment of corn stover to enhance enzyme saccharification. Ind. Crop. Prod. **37**, 352–357 (2012)

Partitioning of Nitralase Enzyme from *Pseudomonas putida* in Polymer/Salt Aqueous Two Phase System

V. Lokesh Ramana, Iyyaswami Regupathi, B.S. Rashmi and S. Nainegali Basavaraj

1 Introduction

Nitriles are organic compounds with –CN as their functional group. Nitriles are important intermediates in the synthesis of amides, amines, esters, carboxylic acid, ketones and heterocyclic compounds (Gong et al. 2012). These are extensively used in chemical industry for the production of solvents, drug intermediates, pharmaceuticals and herbicides. Nitrilase (EC 3.5.5.1) is an enzyme, which can be used as a green catalyst for the hydrolysis of the nitriles to their respective carboxylic acids and ammonia, it is of great importance because of their ability to convert toxic nitriles to high value acids and amides under mild conditions without affecting other reactive groups (Kaul et al. 2004).

Various kinds of purification techniques like salt precipitation, chromatography, gel filtration have been used to purify nitrilase (Banerjee et al. 2006). All these methods were expensive and difficult to scale up at industrial level. Overall design of a process must be consider in order to scale up a technique which will reduce the downstream costs and operational steps. Aqueous two phase extraction is one of the selective extraction method. ATPS has several advantages over other techniques such as system consists of 80–90 % of water which reduces the degradability of target biomolecules, high biocompatibility, good resolution, simple process scale-up and low interfacial tension. In the present study, the partition behavior of Nitralase produced from *Pseudomonas putida* in polymer/salt systems with different salts was studied in detail. Binodal curves were constructed and tie lines were

V. Lokesh Ramana · I. Regupathi (✉) · B.S. Rashmi · S.N. Basavaraj
Department of Chemical Engineering, National Institute of Technology Karnataka, Surathkal, Mangalore 575025, India
e-mail: regupathi@nitk.ac.in

Prasanna B.D. et al. (eds.), *Biotechnology and Biochemical Engineering*,
DOI 10.1007/978-981-10-1920-3_10

generated for PEG/Salt systems and equilibrium phase compositions of the two-phase systems were analysed using Othmer-Tobias and Bancroft equations. The partitioning behavior of Nitralase was investigated by varying system parameters.

2 Materials and Methods

Pseudomonas putida culture was grown on media described by (Banerjee et al. 2006) and incubated for 12 h at 30 °C in an orbital shaker (150 rpm) which was subjected to homogenization followed by centrifugation (12,000 rpm at 4 °C) and for the partitioning studies supernatant was used. To study the partition behavior of enzyme, crude extract of 30 % (w/w) was added to different composition of ATPS. Total protein and Nitralase activity in Top and bottom phase samples were analysed by Bradford assay (Bradford 1976) and nessler method (Banerjee et al. 2006) respectively.

Cloud point method was employed to obtain the binodal curves (Bolar et al. 2013). Tie lines were determined (Eq. 2) by preparing a suitable biphasic system of salt and polyethylene glycol of total weight of 10 g. Systems were allowed to equilibrate at 25 °C for 24 h and the separated individual phases were analyzed for the concentration of salt and polymer by using flame photometer and digital refractometer respectively. The concentration of polymer as well as salt influences the Refractive index of the solution. For dilute aqueous solution containing a polymer and salt, the relation between refractive index (ηD) and the mass fraction of polymer (Wp) and salt (Ws) is given by

$$\eta\,D = a_0 + a_1 W_p + a_2 W_s \tag{1}$$

where W_p is the weight fraction of polymer and W_s are the and salt respectively. a_0, a_1, a_2 are the fitting parameters and were presented in Table 1.

Tie line length (TLL),

$$TLL(\text{wt\%}) = \sqrt{(Ws^t - Ws^b)^2 + (Wp^t - Wp^b)^2} \tag{2}$$

where Ws^t and Ws^b are the salt concentration at top and bottom phase (%W/W), Wp^t and Wp^b corresponds to polymer concentration in top and bottom phase (% w/w).

Table 1 Fit parameters for Eq. (1)

ATPS	a_0	a_1	a_2
PEG1000/Na_2SO_4	1.331	0.127	0.155
PEG1000/NaH_2PO_4	1.331	0.144	0.105
PEG1000/$Na_3C_6H_5O_7$	1.331	0.148	0.168
PEG1000/$K_3C_6H_5O_7$	1.331	0.144	0.130

Pseudomonas putida Putrid culture was grown on media described by Banerjee et al. (2006) and incubated for 12 h at 30 °C in an orbital shaker (150 rpm) which was subjected to homogenization followed by centrifugation (12,000 rpm at 4 °C) and supernatant was used for further partitioning studies. To study the partition behavior of enzyme, 30 %(w/w) of crude extract was added to different composition of ATPS. Top and bottom phase samples were analysed for total protein content by Bradford assay (Bradford 1976) and Nitralase activity by measuring the amount of ammonia produced using nessler method using mandelonitrile as a substrate (Banerjee et al. 2006). The important parameters in ATPS which characterize the partitioning behavior of the enzyme like, Activity partitioning coefficient (Ke), Partitioning coefficient of total protein (Kp), Enzyme purification folds and %yield were analyzed at different conditions. All these parameters were find out by using the equations described in Bolar et al. (2013).

3 Results and Discussion

3.1 *Effect of Type of Salt*

The enzyme partition behavior depends on the concentration and type of phase forming components, polymer Molecular weight, pH, volume ratio and equilibrium characteristics. Four different salts (sodium sulphate, sodium dihydrogen phosphate, Trisodium citrate and Tripottasium citrate) effect were investigated on partitioning of Nitralase with PEG1000 concentration and constant 10 %w/w salt concentration. The phase forming ability of all the four salts were evaluated with PEG-1000. Binodal data was generated through cloud point method for PEG 1000-salt combinations except PEG1000/tripottasium citrate which was obtained from Kalaivani et al. (2013). The binodal curves shape and position depends on the type of salt and polymer used. From the binodal curve (Fig. 1) it was known that the concentration of salts required to form two phase are in the sequence of $Na_2SO_4 < Na_3C_6H_5O_7 < NaH_2PO_4 < K_3C_6H_5O_7$. Decreased cationic radius of sodium increased the salting out ability. Two phase region formed was high in case of sodium salts (0.098 nm) due to decreased cationic radius than potassium salt (0.133 nm). It was observed that cations with more negative Gibbs free energies of hydration (ΔGhydation) require less salt concentration to form biphasic system consequently less amount of sodium salts (Na+: −375 kJ/mol) were required to salt out PEG when compared to potassium salt (K+: −304 kJ/mol).

From Fig. 1b it was seen that anionic species (So_4^{2-}, HPo_4^{2-}, acetate) which has high charge density are more effective and consequently affected the Nitralase partition. Thus sodium sulfate with high sulfate charge density and high salting out ability expelled nitralase to a top phase which increases the hydrophobic interaction

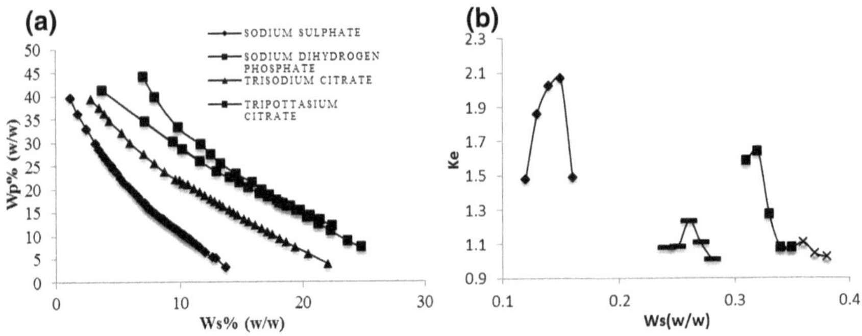

Fig. 1 **a** Binodal curves for PEG1000/salt system; **b** Partitioning coefficient of nitrilase with different salts [Sodium sulphate (*filled diamond*), Sodium dihydrogen phosphate (*filled square*), Trisodium citrate (–) and Tri potassium citrate (x)]

between PEG and enzyme giving highest partition K_e when compared to other salts. The best Ke of 2.066 and yield of 67.38 % was obtained for the salt sodium sulfate.

3.2 Effect of PEG and Salt Concentration

The selection of the polymer and its concentration is one of the critical factor to achieve maximum partitioning in an ATPS. Different PEG MW (1000, 2000, 4000, 6000 and 8000) were investigated on partition coefficient of enzyme.

Partitioning was carried out with five different MWs of PEG and with different PEG (12–17 %) concentration with constant 10 % sodium sulfate. It was observed from Fig. 2a that increases in molecular weight and concentration of the PEG decreases the partition coefficient of an enzyme. This trend was observed due to volume exclusion effect and increased viscosity and interfacial tension. At Molecular weight of PEG-1000 and concentration of 15 % with constant 10 % sodium sulfate maximum ke of 2.066 and yield of 67.39 % was achieved. Mehrnoush et al. (2011) observed similar findings in pectinase partition from mango waste.

Effect of concentration of salt on partitioning of nitrilase was studied by keeping the PEG1000 (15 %w/w) concentration constant and varying the salt concentration from 10–14 %w/w. From Fig. 2b, it was noticed that on increasing the salt concentration above 10 % the partitioning of total protein increased from 1.426 to 2.56 whereas the partitioning coefficient of the nitrilase enzyme decreased. This could be due to the salting out effect of the salt which resulted in expelling the contaminating protein to the top phase. Yucekan et al. (2011) reported similar results when partitioning invertase in PEG-3000 and sodium sulphate system. Bolar et al. (2013) also observed identical findings for the enzyme L-glutaminase from *Zacharomycex rouxii*.

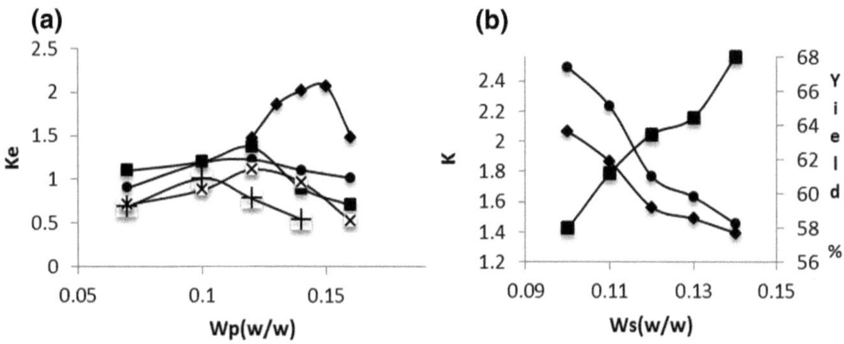

Fig. 2 **a** Effect of PEG on nitrilase partitioning coefficient at different PEG [MW—1000 (*filled diamond*), 2000 (*filled square*), 4000 (*filled circle*), 6000 (x) and 8000 (+)]. **b** Effect of concentration of salt on Ke (*filled diamond*), Kp (*filled square*) and yield % (*filled circle*)

3.3 *Effect of Tie Line Length (TLL)*

Tie line length describes the phase forming components concentration present in the top and bottom phase. The partition coefficient of nitrilase tends to increase up to a TLL% of 35.09 and above this TLL there was precipitation observed at the interphase for PEG1000/sodium sulfate system (Fig. 3a).Increase in TLL increases the phase forming components concentration in the respective phases. Hydrophobic interactions between PEG and protein increases with increase in PEG thus reduces the free volume of salt phase which results in effective salting out which moves the protein towards top phase thus increase in partition coefficient of an enzyme. At very high TLL the physical properties like density and viscosity of the system increases and also available free volume for target biomolecule to accommodate in given phases decreases thus enzyme partitioned to interphase above 35.09 %. Very low TLL fails to give a sharp resolution between the phases. Mehrnoush et al. (2013) reported findings in serine protease recovery from mango waste where the maximum k was observed at TLL of 28.5 and decreases above that, this result was in agreement with glutaminase partition reported by Bolar et al. (2013).

3.4 *Effect of PH*

Change in pH alters the composition of ions in the phases and changes the properties of a biomolecule. Different range of pH from 5.5 to 10 was analyzed in system of 15 %w/w PEG1000 and 10 % w/w sodium sulphate on Nitralase partition. It was noticed from Fig. 4a that the Ke and Yield % of nitrilase increased up to pH 7.5 and decreased above that. This could be due to negative charge acquired by enzyme above its isoelectric point which increases its affinity towards PEG and

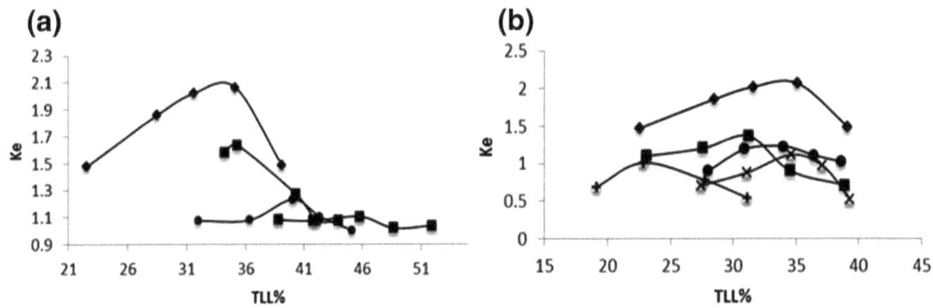

Fig. 3 **a** Effect of TLL on Ke with PEG -1000/salt (Sodium sulphate (*filled diamond*), Sodium digydrogenphosphate (*filled square*), Trisodiumcitrate (*filled circle)* Tripotassium citrate (x). **b** Effect of TLL on Ke with PEG-1000 (*filled diamond*), 2000 (*filled square*), 4000 (*filled circle*), 6000 (x) and 8000 (+) with Sodium sulphate

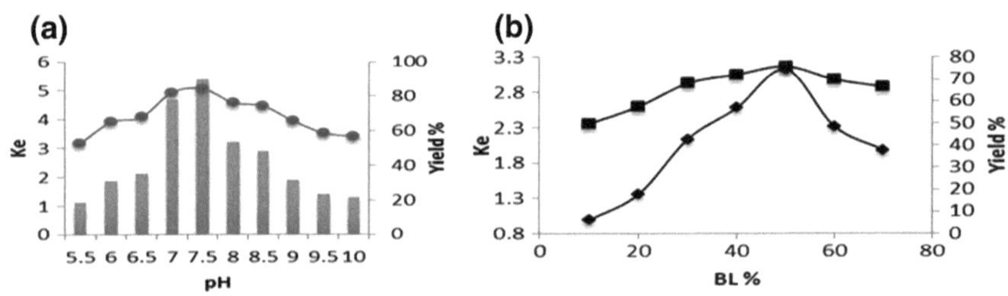

Fig. 4 **a** Effect of pH on Ke (*filled square*) and yield % (*filled circle*); **b** Effect of broth loading (BL)

leads to high partition. Similar results were reported by yang et al. (2008) in partitioning of plant esterase in PEG1000/NaH_2PO_4 and (Kavakcioglu et al. 2013) in partitioning of catalase from *Phanerochaete chrysosporium.* Maximum Ke of 5.412 and yield % 84.40 % was obtained at pH 7.5

3.5 *Effect of Broth Loading and Phase Volume Ratio (Vr)*

Effect of broth loading was studied by adding the fermentation broth from 10 to 70 %w/w to the PEG1000 (15 %w/w) and sodium sulfate (10 %) ATPS. From Fig. 4b it was observed that up to 50 % of broth loading K_e increases and decreases with further increase. When broth loading % increases above certain value enzyme as well as the contaminant protein concentration increases. These impurities compete with nitrilase in the top phase for the free volume and exclude it from the top phase and results in poor K_e. Similar results were obtained by in purification of lipase from recombinant *E. coli* BL21 and (Khayati et al. (2013). At Broth loading of 50 % a purification factor of 2.525 and yield of 75.75 % and ke of 3.125 was obtained for nitrilase from *Pseudomonas putida.*

The effect of volume ratio on nitralase partition was studied at TLL 35.09 % by employing different range of Vr from 0.5 to 2 at pH 7.5 and 50 % Broth loading. The equilibrium composition of phase forming salts remains constant irrespective of the volume ratio. It was observed that purification factor increases up to a volume ratio of 1.5 and decreases with further increase this was observed because, at high volume ratio saturation limit was attained by Particular phases which tends the proteins to precipitate at interface. The optimum volume ratio is 1.5 with Ke 5.482, yield 89.15 % purification fold 2.46.

4 Conclusions

The partition study of Nitralase from *Pseudomonas putida* was successfully. The phase diagrams for four salts (sodium sulphate, sodium dihydrogen phosphate, Trisodium citrate and tripottasium citrate) with PEG 1000 was developed. The system comprising of PEG1000/sodium sulfate gave maximum two phase region and yield higher partition coefficient in top phase. The effect of phase forming components, TLL, broth loading, concentration of phase component, polymer molecular weight and pH was effectively studied. It was found that increase in molecular weight of the PEG decreases the partition coefficient of the enzyme. TLL of 35.02 %, Broth loading of 50 %, pH 7.5 and volume ratio of 1.5 exhibit maximum partition coefficient for Nitralase enzyme. It was investigated by present study that PEG/sodium sulfate system can be effectively employed for the selective extraction of Nitralase enzyme from fermentation broth.

References

Banerjee, A., Kaul, P., Banerjee, U.C.: Purification and characterization of an enantioselective arylacetonitrilase from Pseudomonas putida. Arch Microbial, **184**:407–418 (2006)

Bolar, S., Belur, P.D., Iyyaswami, R.: Partitioning studies of glutaminase in polyethylene glycol and salt based Aqueous two-phase system. Chem. Eng. Technol. **8**, 1378–1386 (2013)

Bradford, M.M.: A rapid and sensitive method for the quantitation of microgram quantities of protein utilizing the principle of protein-dye binding. Anal. Biochem. **72**, 248–254 (1976)

Gong, J.S., Lu, Z.M., Li, H., Shi, J.S., Zhou, Z.M., Xu, Z.H.: Nitrilases in nitrile biocatalysis: recent progress and forthcoming research. Microb. Cell Fact. **11**, 142 (2012)

Kalaivani, S., Regupathi, I.: Partitioning studies of a-lactalbumin in environmental friendly poly (ethylene glycol)—citrate salt aqueous two phase systems. Bioprocess Biosyst. Eng. **90**, 342–350 (2013)

Kaul, P., Banerjee, A., Mayilraj, S., Banerjee, U.C.: Screening for enantio selective nitrilase: kinetic resolution of racemic mandelonitrile to (R)-(−)—mandelic acid by new bacterial isolates. Tetrahedron: Asymmetry, **15**:207–211 (2004)

Kavakçıoğlu, B., Tarhan, L.: Initial purification of catalase from *Phanerochaete chrysosporium* by partitioning in poly(ethylene glycol)/salt aqueous two phase systems. Sep. Purif. Technol. **105**, 8–14 (2013)
Khayati, G., Alizadeh, S.: Extraction of lipase from *Rhodotorula glutinis* fermentation culture by aqueous two-phase partitioning. Fluid Phase Equilib. **13**(353), 132–134 (2013)
Mehrnoush, A., Sarker, M.Z.I., Mustafa, S., Yazid, A.M.M.: Direct purification of pectinase from mango (*Mangifera Indican Cv. Chokanan*) peel using a PEG/salt-based aqueous two phase system. Molecules **16**: 8419–8427 (2011)
Mehrnoush, A., Yazid, A.M.M.: Purification and recovery of serine protease from mango (Mangifera indica cv.Chokanan) waste using aqueous two-phase system: potential low cost of enzyme and purification method. J. Food, Agri. Environ. **11**(3 & 4): 4 0–4 6 (2013)
Yang, S., Huang, Z., Jiang, Z., Li, L.: Partition and purification of a thermostable xylanase produced by Paecilomyces thermophila in solid-state fermentation using aqueous two-phase systems. Process Biochem. **43**(1), 56–61 (2008)
Yücekan, I., Önal, S.: Partitioning of invertase from tomato in poly(ethylene glycol)/sodium sulfate aqueous two-phase systems. Process Biochem, **46**, 226–232 (2011)

Aqueous Two-Phase (Acetonitrile–Potassium Citrate) Partitioning of Bovine Serum Albumin: Equilibrium and Application Studies

Badekar Sagar Dilip, Regupathi Iyyaswami, Basavaraj S. Nainegali and B.S. Rashmi

1 Introduction

Aqueous two-phase extraction systems (ATPE) are formed by mixing distinctive hydrophilic solvents, immiscible polymers and salts in some ranges of concentration and are highly used in the separation and purification of biomolecules (Liu et al. 2012) such as proteins (Asenjo and Andrews 2011), nucleic acids (Luechau et al. 2009), enzymes (Mehrnoush et al. 2012), flavor compounds (Cardoso et al. 2013), antioxidants and antibiotics (Azevedo et al. 2009) and low molecular weight products. Partitioning of biomolecules in ATPE systems is affected by many factors, including molecular weight/size of polymer, concentration of solvent, salt type and its ionic strength (Brooks et al. 1985), addition of salts, such as NaCl, the degree of pH, temperature and affinity of the macromolecule towards polymer or solvent (Alberttson 1986; Ratanapongleka 2010). The primary point of interest of ATPE strategy is that it generously decreases the number of primary steps of downstream processing mainly partial purification is coordinated in one unit. It can be used in combination of other separation methods such as liquid chromatography (Alberttson 1986), packed column (Rosa et al. 2012), magnetic particle adsorption (Gai et al. 2011). However some difficulties during the scale-up of the ATPS in the industry was reported, especially polyethylene glycol (PEG)/salt system causing corrosion of equipment and precipitation of target product.

In the past decades, ATPS have shown capable to be created by two polymers [dextran/polyethylene glycol (Grolbmann et al. 1995)] or by a polymer-salt combination [polypropylene glycol/$(NH_4)_2SO_4$, $MgSO_4$, KCl or KCH_3CO_2] (Zhao et al. 2011) can be labeled as "traditional systems". In recent times, other compounds have been successfully used in the replacement of the traditional con-

B.S. Dilip · R. Iyyaswami (✉) · B.S. Nainegali · B.S. Rashmi
Department of Chemical Engineering, National Institute of Technology Karnataka, Surathkal, Mangalore 575025, Karnataka, India
e-mail: regupathi@nitk.ac.in

Prasanna B.D. et al. (eds.), *Biotechnology and Biochemical Engineering*,
DOI 10.1007/978-981-10-1920-3_11

stituents, such as the pairs alcohol–salt (Reis et al. 2012), ionic liquid–salt (Neves et al. 2009), ionic liquid–polymer (Pereira et al. 2013), and ionic liquid–carbohydrate (Freire et al. 2011). Recently, pioneering ATPS based on acetonitrile (ACN) and sugars have also been reported (Wang et al. 2008).

Acetonitrile (CH_3CN) is an aprotic organic solvent miscible with water in the whole composition range and its molecules do not strongly interact with themselves leaving a hydrogen bond network formed by water (Takamuku et al. 1998). Acetonitrile is a by-product from the manufacture of acrylonitrile (Pollak et al. 2000) widely used in pesticide, rubber, perfume, pharmaceutical industries and also used as mobile phase in reverse phase high performance liquid chromatography in separation and purification processes (Taha et al. 2012). Aiming to find novel phase-forming components to create ATPS, this work found innovative ATPS formed by acetonitrile and potassium citrate ($C_6H_5K_3O_7$) as a salt and water. The corresponding phase diagrams, tie-lines and tie-line lengths were determined at 303.15 K. Moreover, to investigate the extractive performance of this novel system and used in the partitioning of BSA protein.

2 Materials and Methods

2.1 *Materials*

All compounds were purchased from Sigma-Aldrich: acetonitrile (HPLC grade with a purity of 99.99 %), Potassium citrate (>99.5 % pure), BSA (>99 % pure).

2.2 *Phase Diagrams and Tie-Lines*

The binodal curve was determined for acetonitrile and potassium citrate by cloud point titration method at temperature 303.15 K and atmospheric pressure. Liquid-liquid equilibrium experiments were carried out by gravimetric method to determine the tie-lines. The mass of the bottom phase was determined from the volume of bottom phase and density of bottom phase and mass of the top phase was obtained by the subtraction method. Then the mass fractions of salts in the top and bottom phases were determined by flame photometer. Refractive index method was used for determination of mass fraction of acetonitrile in both phases using the following relation between refractive index of solution, (n_D), and the mass fractions of acetonitrile (w_1) and $C_6H_5K_3O_7$ (w_2) (Zafarani-Moattar et al. 2013)

$$n_D = a_0 + a_1 w_1 + a_2 w_2 \tag{1}$$

Here, a_0 is the refractive index of pure water for which obtained value of 1.33192 at T = 303.15 K. Two constants a_1 and a_2 corresponding to acetonitrile and potassium citrate respectively are obtained by using solver in excel and linear calibration plot of refractive index of the solution are obtained (a_1 and a_2 are 0.0299 and 0.1652, respectively).

Also the slope of the tie-line and tie-line length at different compositions were calculated using the Eqs. 2 and 3.

$$S = (W_1^t - W_1^b)/(W_2^t - W_2^b) \quad (2)$$

$$TLL = [(W_1^t - W_1^b)^2 + (W_2^t - W_2^b)^2]^{0.5} \quad (3)$$

where, w_1^t and w_2^t represent the equilibrium compositions of acetonitrile and potassium citrate mass fractions in the top phase, similarly w_1^b and w_2^b in bottom phase respectively.

2.3 *Partitioning of Bovine Serum Albumin*

The partitioning of BSA in acetonitrile and potassium citrate was performed by adding 1 mg/mL in the initial aqueous solution along with the phase forming components. After complete mixing for 5 min, each tube was placed in a thermostatic bath at 303.15 K for about 18 h for phase separation. The volume of each phase was initially measured and both phases were further separated for the quantification of BSA and for determination of their density, viscosity and pH values. The concentration of BSA at each aqueous phase was quantified through Bradford Assay. The partition coefficient and recovery of BSA was determined according to Eqs. 4 and 5.

$$K_{BSA} = \frac{C_T}{C_B} \quad (4)$$

$$R_T = \frac{C_T}{C_T + C_B} * 100 \quad (5)$$

3 Results and Discussion

3.1 *Phase Diagram Data and Correlation*

The experimental binodal data of the Acetonitrile + potassium citrate + water system at (303.15 K) was correlated using Eq. 6 (Wang et al. 2009).

Table 1 Values of parameters of Eq. 6 for the Acetonitrile + $C_6H_5K_3O_7$ + H_2O system at temperature T = 303.15 K

ATPS (303.15)	a	b	c	SD	R^2
Acetonitrile + $C_6H_5K_3O_7$	0.9766	−2.9335	2.0319	0.0214	0.9984

$$\mathbf{w_1} = (\mathbf{a} + \mathbf{b}\ \mathbf{w_2^{0.5}} + \mathbf{c}\ \mathbf{w_2}) \quad (6)$$

where, w_1 and w_2 are the mass fractions of acetonitrile and potassium citrate, and the coefficients a, b and c, corresponding correlation coefficient (R^2) and corresponding standard deviations (SD) of Eq. 6 for the investigated system are given in Table 1.

$$sd = \sum_{i=1}^{n}[\left(W^1_{cal} - W^1_{exp}\right)^2/n]^{0.5}$$

Where, n represent the number of binodal data, W^1_{exp} is the experimental mass fraction of acetonitrile and W^1_{cal} is the corresponding data calculated. The binodal curve which is closer to the coordinate axis indicates the easier formation of the phases (Guoa et al. 2012). Here, Potassium citrate shows higher tendency of phase separation besides that, the conclusion was also explained by the effective excluded volume (EEV) theory of salts (Han et al. 2013) where V* of 360.416 was obtained for our system with Eq. 7.

$$\mathbf{ln}\left(\mathbf{v}^* \frac{w_2}{m_2}\right) + \left(\mathbf{v}^* \frac{w_1}{m_1}\right) = \mathbf{0} \quad (7)$$

3.2 *Liquid-Liquid Equilibrium Data and Correlation*

The experimental tie-line data of the acetonitrile + potassium citrate + H_2O systems at 303.15 K are shown in Table 2 and Fig. 1. The Othmer–Tobias equation (Eq. 8)

Table 2 Tie-line data for the acetonitrile + $C_6H_5K_3O_7$ + H_2O systems at temperature T = 303.15 K

ATPE at 303.15 K	Total system		Top system		Bottom system		TLL (wt %)	Slope
	100 W_1	100 W_2	100 W_1^t	100 W_2^t	100 W_1^b	100 W_2^b		
Acetonitrile + $C_6H_5K_3O_7$	39	7	48.118	3.300	19.189	13.964	30.83	−2.712
	41	7.4	52.222	2.756	17.960	15.415	36.52	−2.706
	43	7.8	56.193	2.357	17.031	16.721	41.71	−2.726
	45	8.2	60.063	1.958	16.236	17.990	46.66	−2.733
	47	8.6	63.199	1.523	15.840	19.187	50.54	−2.681

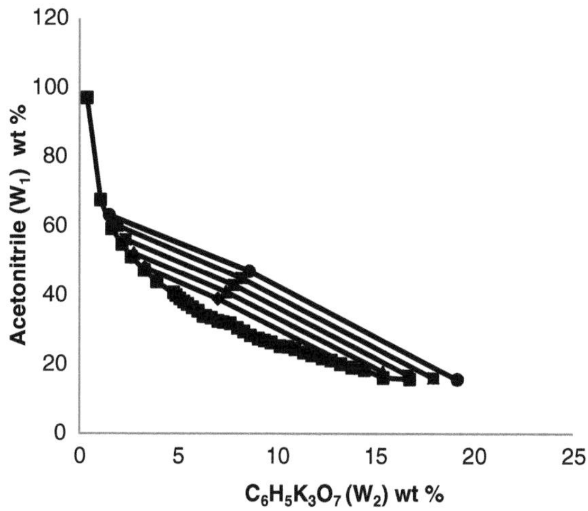

Fig. 1 Tie-lines at T = 303.15 K acetonitrile/potassium citrate ($C_6H_5K_3O_7$) ATPS

Table 3 Values of parameters of Eq. 8 for acetonitrile + $C_6H_5K_3O_7$ + H_2O systems at temperature 303.15 K

ATPS at 303.15 K	K_1	n	R^2	SD
Acetonitrile + $C_6H_5K_3O_7$	0.16980	0.89981	0.99977	0.11325

was used for correlating the LLE data of acetonitrile—$C_6H_5K_3O_7$ ATPS (Zhao et al. 2011).

$$\frac{(1 - W_1^t)}{W_1^t} = K_1 \left(\frac{(1 - W_2^b)}{W_2^b} \right)^n \tag{8}$$

Here, K_1 and n are fitting parameters. As per the values of R^2 and SD in Table 3, it can be concluded that Othmer–Tobias equation can be satisfactorily applied for correlating the tie-line data of the studied system.

3.3 *Partitioning of Bovine Serum Albumin*

Acetonitrile-potassium citrate system showed partition coefficient of BSA more than 1 and showed the preferential interaction of BSA towards the acetonitrile-rich phase. The partition coefficient increased with increase in tie line length of system (Fig. 2a) and the hydration capacity of the salts prompts to an exclusion impact of the BSA towards the acetonitrile-rich phase. Effect of acetonitrile concentration on partitioning of BSA was investigated at salt concentration of 7 wt% with varying

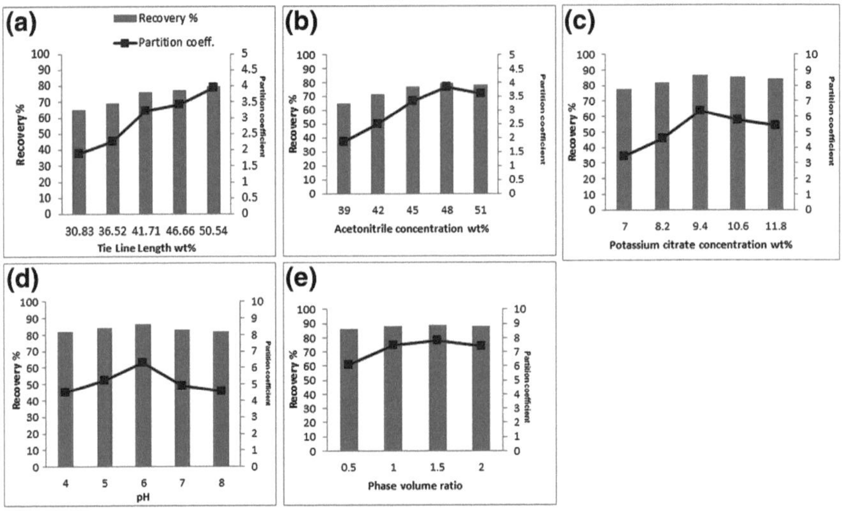

Fig. 2 Effect of variables on top phase recovery % and the total BSA partition coefficient in acetonitrile—$C_6H_5K_3O_7$ system. **a** Effect of TLL. **b** Effect of Acetonitrile concentration. **c** Effect of Potassium citrate concentration. **d** Effect of pH. **e** Effect of phase volume ratio

acetonitrile concentration in the range of 39–51 wt%. The maximum partition coefficient was obtained at acetonitrile concentration of 48 wt% with a maximum k of 3.83(Fig. 2b). As concentration of acetonitrile increases, the free volume available for the protein to partition into the top phase decreases causing the protein to move into bottom phase as observed for acetonitrile concentration higher than 48 wt%. Partition coefficient increased up to salt concentration 9.4 wt% with constant acetonitrile concentration. Maximum k value obtained for BSA in this system is 6.38 (Fig. 2c). Finally the system composition 48 wt% acetonitrile and 9.4 wt% potassium citrate having TLL 53 wt% showed highest top phase BSA recovery of 86.45 %. In aqueous two phase system stronger phase separating compounds improve the migration of the biomolecule for the opposite phase. Potassium citrate showed good salting out capacity which is best for biomolecule migration in opposite phase which is confirmed with BSA protein in our studies. After particular concentration of salt the partition coefficient decreases, this may be because of denaturation of biomolecule in presence of higher salt concentration and accumulation of protein in the interphase (Kalaivani et al. 2012).

The ratio of charged species present in the system and also the net charge of the biomolecule are known to vary with the system pH. Above isoelectric point, net charge of the protein is negative and further increase in negativity depends on the presence of charged amino acids on the protein surface. The variation of partition coefficient with pH was analyzed for TL of 48 wt% acetonitrile—9.4 wt% potassium citrate system and is reported in (Fig. 2d). Partition coefficient of protein was observed to increase from system pH 4 to 6 and again it was decreasing at system pH 7 and 8. The pI of BSA is 4.7 above which it is negatively charge. Acetonitrile is also negatively charged which means that BSA must show maximum partitioning below pH 4.7, but as per the results obtained we can find maximum partitioning of BSA at pH 6 showing that partitioning of BSA is majorly affected by the salting out effect and not the charge of the molecule.

Choice of a proper volume proportion is also important to achieve maximal recovery of protein. The equilibrium compositions of the phase forming components (Acetonitrile and salt) are constant regardless of the volume ratio along a TLL at constant system pH 6. Considering both the recovery and partition coefficient (Fig. 2e), it was concluded that the volume proportion of 1.5 at a TLL of 53 wt% and system pH 6 is suitable to obtain a higher partition coefficient and recovery of BSA was 7.78 and 88.61 % respectively.

4 Conclusions

ATPS can be formed by acetonitrile and potassium citrate at specific concentrations in aqueous media. The phase diagrams, tie-lines and tieline lengths were determined at 303.15 K and atmospheric pressure. Potassium citrate showed higher tendency of phase separation, the conclusion was also explained by the effective excluded volume (EEV) theory of salts. The liquid–liquid equilibrium data were predicted through the Othmer–Tobias correlations. Bovine Serum Albumin (BSA) was considered to access the partition ability of the Acetonitrile—potassium citrate system. BSA preferentially partitions for the acetonitrile-rich phase with maximum partition coefficient higher than 7. Moreover, the BSA recovery obtained at optimal conditions was higher than 88 %, supporting the huge potential of this system to be explored in the extraction of the most diverse value added compounds for industrial applications.

References

Alberttson, P.A.: Partition of Cell Particles and Macromolecules. Separation and Purification of Biomolecules, Cell Organelles, Membranes and Cells in Aqueous Polymer Two-phase Systems and Their Use in Biochemical Analysis and Biotechnology. Wiley-Interscience, New York (1986)

Asenjo, J.A., Andrews, B.A.: Aqueous two-phase systems for protein separation: a perspective. J. Chromatogr. A **1218**, 8826–8835 (2011)

Azevedo, A.M., Sommerfeld, S., Mutter, A., Aires-Barros, M.R., et al.: Application of aqueous two-phase systems to antibody purification: a multi stage approach. J. Biotechnol. **139**, 306–313 (2009)

Brooks, E., Sharp, K.A., Fischer, D.: Partitioning in Aqueous Two-Phase Systems: Theory, Methods, Uses and Applications to Biotechnology. Academic Press, New York (1985)

Cardoso, G., Mourao, T., Pereira, F., Freire, M., Fricks, A., Soares, C., Lima, A.: Aqueous two-phase systems based on acetonitrile and carbohydrates and their application to the extraction of vanillin. Sep. Purif. Technol. **104**, 106–113 (2013)

Freire, M.G., Louros, C.L.S., Rebelo, L.P.N., Coutinho, J.A.P.: Aqueous biphasic systems composed of a water-stable ionic liquid + carbohydrates and their applications. Green Chem. **13**, 1536–1545 (2011)

Gai, Q., Qu, F., Zhang, T., Zhang, Y.: Integration of carboxyl modified magnetic particles and aqueous two-phase extraction for selective separation of proteins. Talanta **85**, 304–309 (2011)

Grolbmann, C., Tintinger, R., Zhu, J., Maurer, G.: Aqueous two-phase systems of poly(ethylene glycol) and dextran—experimental results and modeling of thermodynamic properties. Fluid Phase Equilib. **106**, 111–138 (1995)

Guoa, W., Ma, J., Wang, Y., Han, J., Li, Y., Song, S.: Liquid–liquid equilibrium of aqueous two-phase systems composed of hydrophilic alcohols (ethanol/2-propanol/1-propanol) and $MgSO_4$/$ZnSO_4$ at (303.15 and 313.15) K and correlation. Thermochim. Acta. **546**, 8–15 (2012)

Han, J., Wu, Y., Xiang, Y., Wang, Y., Ma, J., Hu, Y.: Liquid–liquid equilibria of hydrophilic alcohol + sodium hydroxide + water systems: Experimental and correlation. Thermochim. Acta **566**, 261–267 (2013)

Kalaivani, S., Srikanth, C.K., Regupathi, I.: Densities and viscosities of binary and ternary mixtures and aqueous two-phase system of poly ethylene glycol 2000 + diammonium hydrogen citrate + water at different temperatures. J. Chem. Eng. Data **57**, 2528–2534 (2012)

Liu, Y., Wu, Z., Zhang, Y., Yuan, H.: Partitioning of biomolecules in aqueous two-phase systems of polyethylene glycol and nonionic surfactant. Biochem. Eng. J. **69**, 93–99 (2012)

Luechau, F., Ling, T.C., Lyddiatt, A.: Primary capture of high molecular weight nucleic acids using aqueous two-phase systems. Sep. Purif. Technol. **66**, 202–207 (2009)

Mehrnoush, A., Mustafa, S., Sarker, M.I., Yazi, A.M.M.: Optimization of serine protease purification from mango (*Mangifera Indica* cv. Chokanan) peel in polyethylene glycol/dextran aqueous two phase system. Int. J. Mol. Sci. **13**, 3636–3649 (2012) (Rosa PAJ)

Neves, C.M.S.S., Ventura, S.P.M., Freire, M.G., Marruchol, M., Coutinho, J.A.P.: Evaluation of cation influence on the formation and extraction capability of ionic liquid-based aqueous biphasic systems. J. Phys. Chem. B **113**, 5194–5199 (2009)

Pereira, J.F.B., Ventura, S.P.M., Silva, F.A., Shahriari, S., Freire, M.G., Coutinho, J.A.P.: Aqueous biphasic systems composed of ionic liquids and polymers: a platform for the purification of biomolecules. Sep. Purif. Technol. **113**, 83–89 (2013)

Pollak, P., Romeder, G., Hagedorn, F., Gelbke, H.P.: Ullmann's Encyclopedia of Industrial Chemistry. Wiley Online Library, New York (2000)

Ratanapongleka, K.: Recovery of biological products in aqueous two phase systems. Int. J. Chem. Eng. Appl. **1**, 191–198 (2010)

Reis, I.A.O., Santos, S.B., Santos, L.A., Oliveira, N., Freire, M.G., Pereira, J.F.B., Ventura, S.P.M., Coutinho, J.A.P., Soares, C.M.F., Lima, A.S.: Increased significance of food wastes: selective recovery of added-value compounds. Food Chem. **135**, 2453–2461 (2012)

Rosa, P.A.J., Azevedo, A.M., Sommerfeld, S., Mutter, M., Bäcker, W., et al.: Continuous aqueous two-phase extraction of human antibodies using a packed column. J. Chromatogr. B **880**, 148–156 (2012)

Takamuku, T., Tabata, M., Yamaguchi, A., Nishimoto, J., Kumamoto, M., Wakita, H., Yamaguchi, T.: Liquid structure of acetonitrile-water by X-ray diffraction and infrared spectroscopy. J. Phys. Chem. B **102**, 8880–8888 (1998)

Taha, M., Teng, H.L., Lee, M.J.: Phase diagrams of acetonitrile or (acetone + water + EPPS) buffer phase separation systems at 298.15 K and quantum chemical modeling. J. Chem. Thermodyn. **54**, 134–141 (2012)

Wang, B., Ezejias, T., Feng, H., Blaschek, H.: Sugaring-out: a novel phase separation on extraction system. Chem. Eng. Sci. **63**, 25958–26000 (2008)

Wang, Y., Hu, S., Yan, Y., Guan, W.: Liquid-liquid equilibrium of potassium/sodium carbonate + 2-propanol/ethanol + water aqueous two-phase systems and correlation at 298.15 K. CALPHAD: Comput. Coupling Phase Diagrams Thermochem. **33**, 726–731 (2009)

Zafarani-Moattar, M., Hosseinpour-Hashemi, V., Banisaeid, S., Beirami, M.: The study of phase behavior of aqueous 1-propanol/2-propanol/2-butanol/2-methyl-2-propanol systems in the presence of disodium tartrate or disodium succinate at T = 298.15 K. Fluid Phase Equilib. **338**, 37–45 (2013)

Zhao, X., Xie, X., Yan, Y.: Liquid–liquid equilibrium of aqueous two-phase systems containing poly (propylene glycol) and salt ((NH4)2SO4, MgSO4, KCl, and KAc): experiment and correlation. Thermochim. Acta **516**, 46–51 (2011)

Mixed Surfactant Based Reverse Micelle Extraction of Lactose Peroxidase from Whey

Sivananth Murugesan, Prudhvi Ambakam, Akshay Naveen, Aarathi Makkada, Nithin Solanki and Regupathi Iyyaswami

1 Introduction

Global production of liquid whey from cheese and casein amounted be over billion to trillion tonnes with respect to increasing demand in the dairy industry (Smithers 2008). β-lactoglobulin, α-lactalbumin, bovine serum albumin and immunoglobulins (IgG) are the major proteins found in whey while LPO and Lactoferrin remain as minor constituents of whey (Bottomley et al. 1990; De Wit 1998). Most of the proteins found in the whey possess health benefits and are directly or indirectly related to the functioning of the immune system. Lactoperoxidase (E.C. 1.11.1.7), LPO as they are commonly referred to, constitute about 30 mg/L or about 1 % of cheese whey that are generated from the Dairy industries. LPO are oxidoreductase enzymes that are found to function in the mammary glands and acts as potent antimicrobial agent in the intestinal tract of the new born infants (Pruitt et al. 1985; Sharma et al. 2013). They have also been employed extensively as food preservants, in cosmetics, and dental applications. Recent studies have been reported on successful usage of LPO towards tumor cells and LPO have been reported to activate immune response—macrophage that act upon the tumor cells (Lefkowitz et al. 1990; Takeshi et al. 1996). Separation of these minor protein fractions from whey has been a difficult task owing to the high concentration of β-lactoglobulin, α-lactalbumin that interferes the separation process. Previous reports on purification of LPO elaborate the usage of different types of chromatography—affinity (Ozdemir et al. 2001), Ion exchange (Atasever et al. 2013) and Hydrophobic interaction chromatography (Langbakk et al. 1984) and reverse micelle extraction (Nandini et al. 2010).

S. Murugesan · P. Ambakam · A. Naveen · A. Makkada · N. Solanki · R. Iyyaswami (✉)
Department of Chemical Engineering, National Institute of Technology Karnataka, Surathkal, Mangalore 575025, India
e-mail: regupathi@nitk.ac.in

Prasanna B.D. et al. (eds.), *Biotechnology and Biochemical Engineering*,
DOI 10.1007/978-981-10-1920-3_12

Liquid–liquid extraction (LLE) has been employed in the selective extraction of such solutes from the feed in general. However, Reverse Micelles based extraction is one such LLE Technique, a versatile, easily scalable and sustainable process that involves the usage of surfactants to form reverse micelle in the presence of an immiscible liquid usually water phase. The solute molecules are solubilized within the reverse micelle, which facilitate the forward extraction and the solutes from reverse micelle are further back extracted into fresh stripping aqueous phase. Water content and solubilization of protein can be altered by changing the aqueous phase pH, with the addition of solvents, co-surfactants and electrolytes (Yazdi 2011; Saha et al. 2015; Mohd-Steapar et al. 2012, 2014). In the present work, the extraction of LPO from whey was studied by employing mixed surfactant based reverse micelles at different conditions.

2 Materials and Methods

Dioctyl sulfosuccinate sodium salt (AOT), cetyltrimethylammonium bromide (CTAB) were purchased from CDH, India. Triton X 100 (TX100) and Isooctane were purchased from Sigma Aldrich, Alcohols—propanol, butanol, pentanol, hexanol, heptanol and octanol were purchased from CDH, India. Dilute HCL, Disodium hydrogen phosphate and sodium dihydrogen phosphate were purchased from Merck, India. Deionized water was used and the experiments were conducted at room temperature unless otherwise specified. Skimmed milk was used for the preparation of whey.

Whey was prepared by altering the pH of skimmed milk to 4.6 with the addition of dilute HCL acid and once the pH was attained, the milk turned into slurry which was centrifuged at 4500 rpm for 10 min and the pellet obtained was discarded while the supernatant was stored at 4 °C for further usage. Sodium phosphate buffer was used as a stripping phase, which was prepared according to standard protocol. The organic phase with surfactant was prepared by dissolving AOT and TX100 in Isooctane, in the presence of Hexanol—23 % (V/V) and Octanol—30 % (V/V) which were added for complete dissolution of the surfactants (different concentrations) in the solvent. The Critical Micelle Concentration (CMC) of the mixed surfactants was obtained from a graph drawn which depicting densities of the solutions and their respective concentrations. Similar experiments were repeated with CTAB/TX100 containing the prescribed amount of Hexanol and Octanol in the stock solution.

LPO assay solution was prepared by diluting 150 μl of 0.09 M hydrogen peroxide to 30 ml of 0.005 M potassium iodide dissolved in 0.033 M sodium phosphate buffer (pH-7.0), 3.0 ml reaction mixture was pipetted into cuvette and incubated at 25 °C for 3–4 min to achieve temperature equilibrium. 100 μl of the enzyme was added and absorbance was recorded at 350 nm for one minute (Morrison and Bayse 1970). Forward Extraction was performed in centrifuge tubes where in equal volumes of the organic phase and the aqueous phase—whey were mixed and the tubes were centrifuged at 3500 rpm for 5 min. From the tube 100 μl of top phase and bottom phase were added to the LPO assay solution and the activity was recorded for a minute at an absorbance of 350 nm. The same was used to calculate the active enzyme units present in 1 ml of whey. The following formulas were used in the calculation of activity of LPO in top and bottom phases.

$$Activity\ (Units/\mathrm{ml}) = \frac{\Delta A\ 350/\mathrm{min}}{(26)} \tag{1}$$

Partition coefficient:

$$(Forward\ Extraction) = \frac{Activity - top\ phase}{Activity - bottom\ phase} \tag{2}$$

$$(Backward\ Extraction) = \frac{Activity - bottom\ phase}{Activity - top\ phase} \tag{3}$$

Whey pH was adjusted and the above protocol was performed to find out the extraction efficiency of the reverse micelles at altered pH; similar experiments were performed for increasing chain length of alcohols from propanol to octanol with varying concentrations respectively. Different Electrolytes were added with the aqueous phase to study the effect of salts on the extraction of LPO from whey to the reverse micelles. Conditions of pH, Alcohol and electrolyte which resulted in maximum partitioning of LPO into the reverse micelle were considered and the forward extraction was performed in bulk. Top phase from the bulk forwards extraction was used for the backward extraction studies, in which one ml of the top phase was used with an equal volume of fresh stripping buffer solution. Different operating conditions were maintained such as pH, alcohols of varying chain length and salts with increasing concentrations, respectively. During the backward extraction process, the top and bottom phases were analyzed for the enzyme activity in the top and bottom phase.

3 Results and Discussion

The CMC value of AOT/TX100/Isooctane and CTAB/TX100/Isooctane solutions were obtained from the density and viscosity measurements and were obtained as 0.1 and 0.01 M, respectively. Both the solutions at their CMC values were used as the organic phase for the extraction of LPO. Effect of pH was studied by varying the pH (7.9–8.7) around the Isoelectric point (pI) of the protein—LPO, since the pI of Bovine LPO has been reported to be 7.9.

From Fig. 1 for AOT/TX100 mixture, it can be inferred that the forward extraction is maximum at a pH of 8.1 while the backward extraction is maximum at a pH of 8.3. As the pH of the solution is maintained above the pI of the protein, the

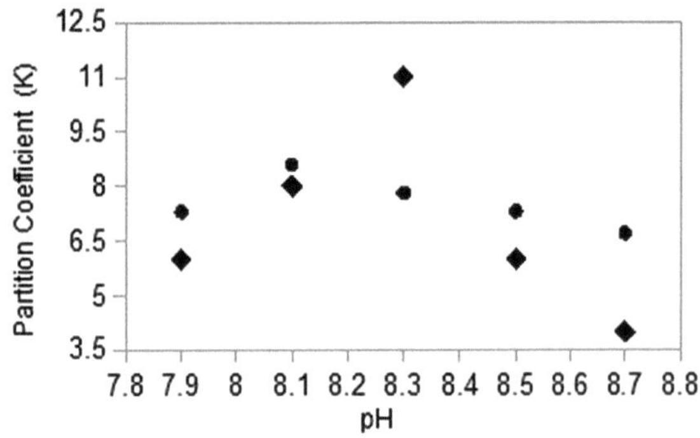

Fig. 1 Effect of pH—AOT/TX100. Forward extraction *filled circle*, backward extraction *filled diamond*

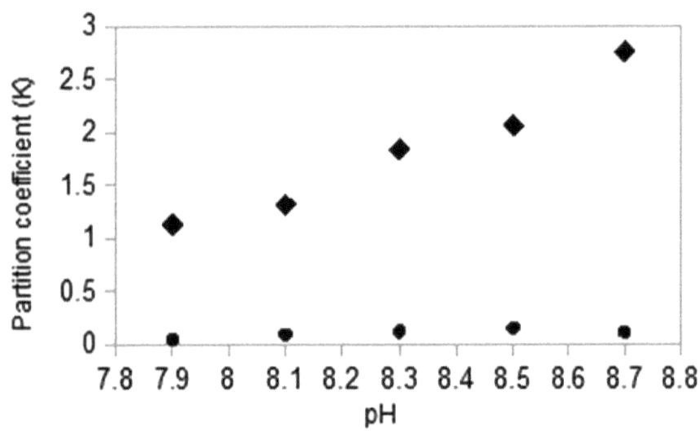

Fig. 2 Effect of pH—CTAB/TX100. Forward extraction *filled circle*, backward extraction *filled diamond*

protein attains a net negative charge and with respect to negatively charge surfactant head group, there should be rejection of protein into the aqueous phase while the vice versa effect noted down might be as a result of enhanced hydrophobic interaction of TX100 present in the mixed surfactant reverse micelle system. With further increase in the pH during backward extraction, the protein attains a strong negative charge which might lead to the expulsion of the same into the fresh stripping aqueous phase. For CTAB/TX100, maximum protein partitioning occurred at 8.5 during forward extraction and backward extraction maximum partitioning from reverse micelle to fresh stripping phase was obtained at a pH of 8.7 as shown in Fig. 2.

Effect of Alcohol on the extraction efficiency of LPO by AOT/TX100 was studied by considering varying chain length of alcohols and different concentrations. It was found from the Fig. 3 that, as the alcohol chain length increases from propanol to butanol the protein partitioning into the top reverse micellar phase increases and with further increase in the chain length butanol to pentanol and heptanol, the extraction efficiency was found to decrease during forward extraction. The said effect is as a result of interaction of alcohol with the surfactant head groups, as the chain length is less the alcohol molecules solubilize in the water core to a smaller extent and also place themselves within the head groups (Chuo et al. 2014; Saran Chaurasiya et al. 2015). As the organic phase contains Hexanol and

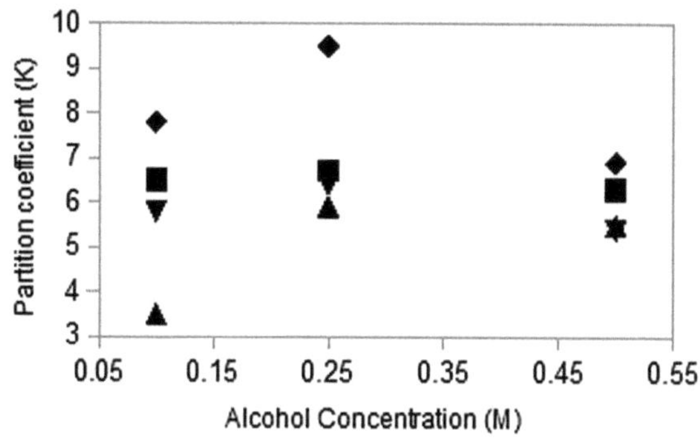

Fig. 3 Effect of alcohol—AOT/TX100. Forward extraction, *filled diamond* Propanol, *filled triangle down* Butanol, *filled triangle up* Pentanol, *filled square* Heptanol

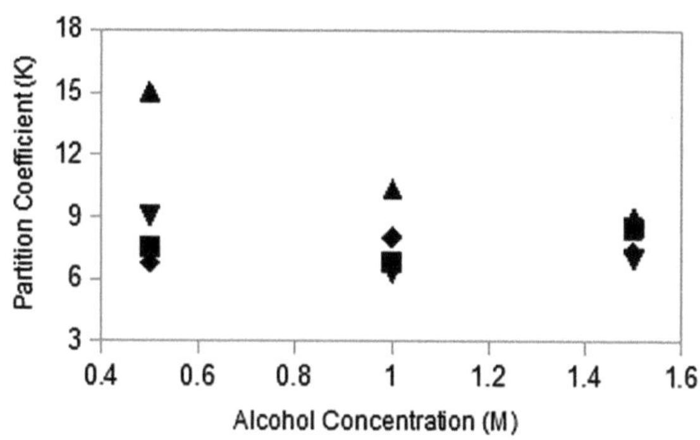

Fig. 4 Effect of alcohol—AOT/TX100. Backward extraction, *filled diamond* Propanol, *filled triangle down* Butanol, *filled triangle up* Pentanol, *filled square* Heptanol

Octanol for proper dissolution of the surfactants in the solvent, the combined effect of higher chain lengths of alcohols leads to breakage of reverse micelles and thereby causing decrease in the mass transfer of LPO from whey to the reverse micelle. During backextraction process, the similar effects were noted as shown in Fig. 4, pentanol leads to breakage of reverse micelle and release of LPO from the reverse micelle water core to the fresh stripping aqueous phase.

When CTAB/TX100, was employed pentanol at a concentration of 0.25 M partitioned LPO from whey to the reverse micelle while heptanol at a concentration of 0.5 M gave maximum LPO back extraction from reverse micelle to the fresh stripping phase as shown in Figs. 5 and 6. Similar effects as seen in AOT/TX100 reverse micellar extraction of LPO were observed and the explanation suits for CTAB/TX100 as well.

Effect of salts on the partitioning of LPO during forward and backward extractions were studied by considering four different salts from Hofmeister series with varying concentrations as shown in the Figs. 7, 8, 9 and 10. It was observed that when AOT/TX100 system was used, ammonium sulphate at a concentration of 0.25 M gave maximum protein partitioning during forward extraction while 0.5 M sodium chloride gave maximum back extraction of the protein. Ammonium ions during forward extraction screen the electrostatic effect between the surfactant head groups and the protein while the sulphate ions hold the protein molecules within the

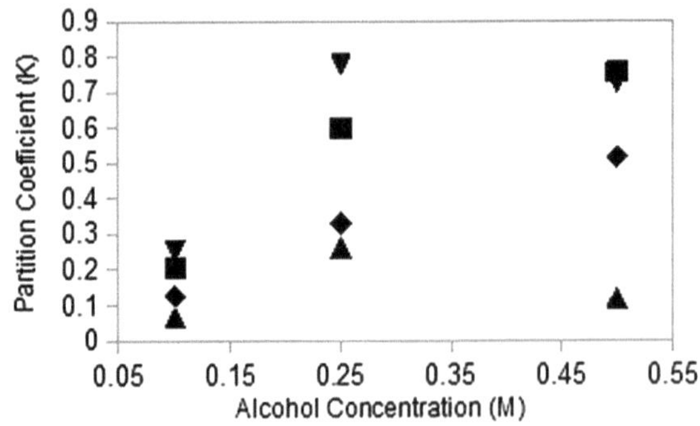

Fig. 5 Effect of alcohol—CTAB/TX100. Forward extraction, *filled square* Propanol, *filled diamond* Butanol, *filled triangle down* Pentanol, *filled triangle up* Heptanol

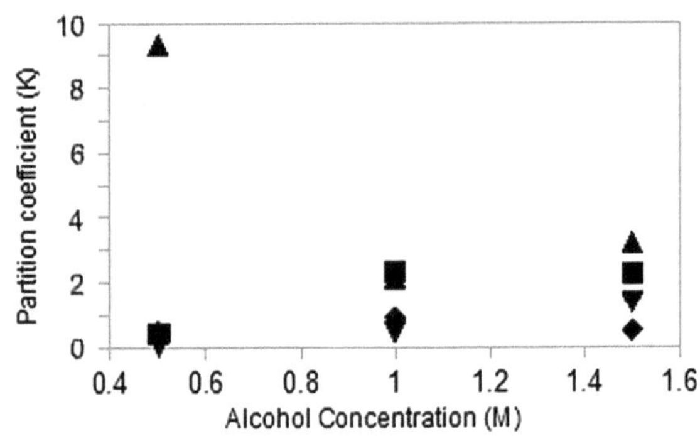

Fig. 6 Effect of alcohol—CTAB/TX100. Backward extraction, *filled diamond* Propanol, *filled square* Butanol, *filled triangle down* pentanol, *filled triangle up* heptanol

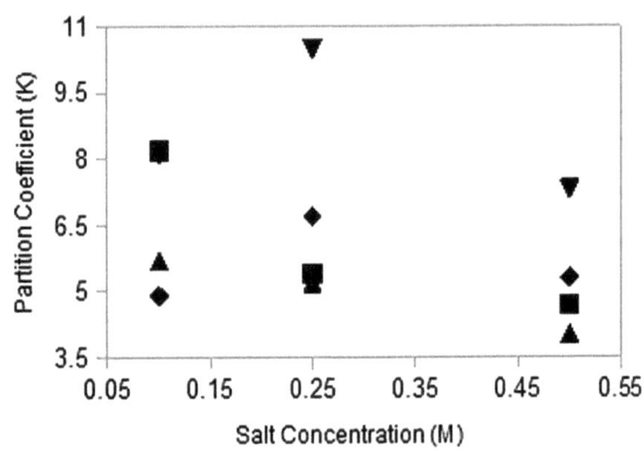

Fig. 7 Effect of salt—AOT/TX100. Forward extraction, *filled diamond* Ammonium Chloride, *filled triangle down* Sodium Chloride, *filled triangle up* Ammonium Sulphate, *filled square* Sodium Sulphate

water core. This effect results in the expulsion of larger proteins like lactalbumin and lactoglobulin while the minor protein concentrations increase in the water core. Sodium chloride resulted in maximum back extraction, sodium ions are smaller in ionic radius compared to ammonium ions that results in enhanced electrostatic attraction between the surfactant head groups and on the protein molecules and as a

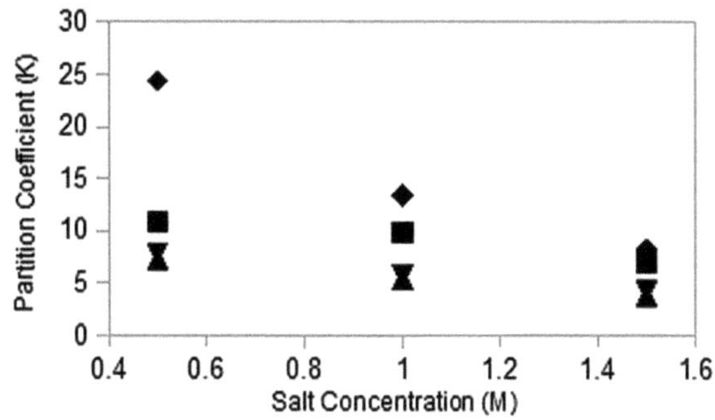

Fig. 8 Effect of salt—AOT/TX100. Backward extraction, *filled diamond* Ammonium Chloride, *filled triangle down* Sodium Chloride, *filled triangle up* Ammonium Sulphate, *filled square* Sodium Sulphate

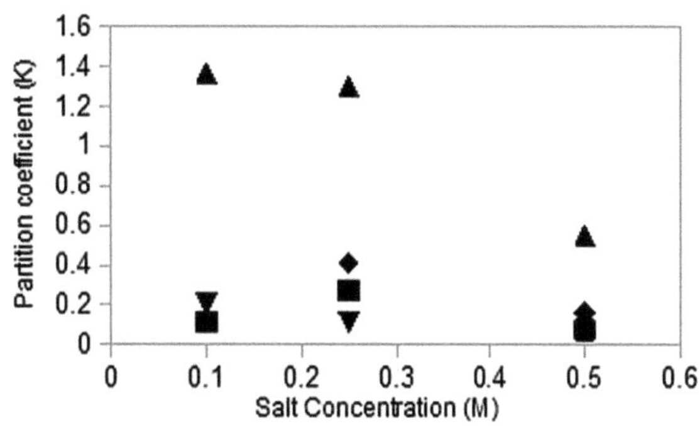

Fig. 9 Effect of salt—CTAB/TX100. Forward extraction, *filled diamond* Ammonium Sulphate, *filled triangle down* Sodium Sulphate, *filled triangle up* Sodium Chloride, *filled square* Ammonium Chloride

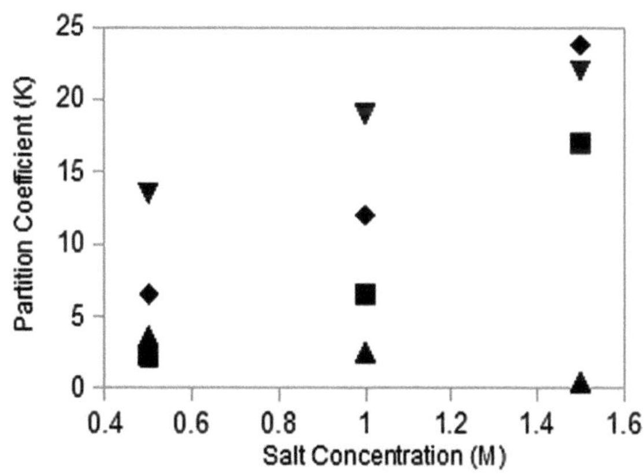

Fig. 10 Effect of salt—CTAB/TX100. Backward extraction, *filled diamond* Ammonium Sulphate, *filled triangle down* Sodium Sulphate, *filled triangle up* Sodium Chloride, *filled square* Ammonium Chloride

result the number of sodium ions in a reverse micelle increases leading to expulsion of water core with the proteins solubilized in it (Yu et al. 2014; Wan et al. 2013).

As CTAB/TX100 reverse micelles were used for the extraction process, Sodium chloride at a concentration of 0.1 M was found to cause maximum partitioning of protein during the forward extraction process; this effect is as a result of electrostatic repulsion of ammonium ions compared to those of sodium ions which are smaller in ionic radius. With increasing concentration of sodium chloride, the protein partitioning was found to decrease; which might be as a result of increase in the chloride ion concentration that leads to protein instability. Ammonium sulphate at a concentration of 1.5 M lead to the disruption of reverse micelles, thereby back extraction of LPO from the reverse micelles to the fresh stripping phase was successfully achieved.

4 Conclusion

AOT/TX100 and CTAB/TX100 mixed surfactant based reverse micelles were formed and their CMC values were determined using density and viscosity measurements and were found to be 0.1 and 0.01 M respectively. Different conditions were tested to observe their effect on extraction efficiency of LPO from whey during the forward and backward extraction steps. It was inferred that maximum protein partitioning occurred when AOT/TX100 reverse micelles were used during the forward extraction process compared to that of CTAB/TX100 reverse micelles. Though the forward extraction of both the systems gave different partitioning values at the end of bulk extraction process, AOT/TX100-10.5, CTAB/TX100-1.37; backward extraction of both the systems resulted in an almost similar back extraction value of AOT/TX100-24.4, CTAB/TX100-23.75. This proves that AOT/TX100 is a better Ionic/Nonionic mixed surfactant system for the extraction of LPO from whey, although back extraction also accounts for the total efficiency of the reverse micelle extraction process.

References

Atasever, A., Ozdemir, H., Gulcin, I., Kufrevioglu, O.I.: One-step purification of LPO from bovine milk by affinity chromatography. Food Chem. **136**(2), 864–870 (2013)

Bottomley, R.C., Evans, M.T.A., Parkinson, C.J.: Whey proteins. In: Food Gels, pp. 435–466. Springer Netherlands (1990)

Chuo, S.C., Mohd-Setapar, S.H., Mohamad-Aziz, S.N., Starov, V.M.: A new method of extraction of amoxicillin using mixed reverse micelles. Colloids Surf. A **460**, 137–144 (2014)

De Wit, J.N.: Nutritional and functional characteristics of whey proteins in food products. J. Dairy Sci. **81**(3), 597–608 (1998)

Langbakk, B., Flatmark, T.: Demonstration and partial purification of LPO from human colostrum. FEBS Lett. **174**(2), 300–303 (1984)

Lefkowitz, D.L., Hsieh, T.C., Mills, K., Castro, A.: Induction of tumor necrosis factor and cytotoxicity by macrophages exposed to LPO and microperoxidase. Life Sci. **47**(8), 703–709 (1990)

Mohd-Setapar, S.H., Mohamad-Aziz, S.N., Harun, N.H., Mohd-Azizi, C.Y.: Review on the extraction of biomolecules by biosurfactant reverse micelles. APCBEE Procedia **3**, 78–83 (2012)

Mohd-Setapar, S.H., Mohamad-Aziz, S.N., Chuong, C.S., Che Yunus, M.A., Ahmad Zaini, M.A., Kamaruddin, M.J.: A review of mixed reverse micelle system for antibiotic recovery. Chem. Eng. Commun. **201**(12), 1664–1685 (2014)

Morrison, M., Bayse, G.S.: Catalysis of iodination by lactoperoxidase. Biochemistry **9**(15), 2995–3000 (1970)

Nandini, K.E., Rastogi, N.K.: Single step purification of LPO from whey involving reverse micelles-assisted extraction and its comparison with reverse micellar extraction. Biotechnol. Prog. **26**(3), 763–771 (2010)

Ozdemir, H., Aygul, I., Küfrevioglu, O.I.: Purification of LPO from bovine milk and investigation of the kinetic properties. Prep. Biochem. Biotechnol. **31**(2), 125–134 (2001)

Pruitt, K.M., Tenovuo, J.O.: The Lactoperoxidase System. Chemistry and Biological Significance, vol. 27. of Immunology Series, Marcel Dekker, New York (1985)

Saha, A., Yadav, R.: Reverse micellar extraction: technological aspects, applications and recent developments. J. Pharm. Res. **9**(2), 145–156 (2015)

Saran Chaurasiya, R., Umesh Hebbar, H., Raghavarao, K.S.M.S.: Recent developments in nanoparticulate based reverse micellar extraction for downstream processing of biomolecules. Curr. Biochem. Eng. **2**(2), 118–134 (2015)

Sharma, S., Singh, A.K., Kaushik, S., Sinha, M., Singh, R.P., Sharma, P., Singh, T.P.: LPO: structural insights into the function, ligand binding and inhibition. Int. J. Biochem. Mol. Biol. **4** (3), 108 (2013)

Smithers, G.W.: Whey and whey proteins—from 'gutter-to-gold'. Int. Dairy J. **18**(7), 695–704 (2008)

Takeshi, O., Mihoko, O., Emiko, H., Naomi, M., Yuko, M., Akiyo, S.: Cytolysis of B-16 melanoma tumor cells mediated by the myeloperoxidase and LPO systems. Biol. Chem. **377** (11), 689–694 (1996)

Wan, J., Li, Q., Zhou, F., Liu, J., Cao, X.: Influence of chaotropes on recovery of trypsin and micellar sizes during reverse micelle extraction. Sep. Purif. Technol. **116**, 307–312 (2013)

Yazdi, A.S.: Surfactant-based extraction methods. TrAC Trends Anal. Chem. **30**(6), 918–929 (2011)

Yu, T., Cao, X.: Effect of chaotropes on lipase back extraction recovery in the process of reverse micellar extraction. Appl. Biochem. Biotechnol. **172**(6), 3287–3296 (2014)

Evaluation of Bio-surfactant on Microbial EOR Using Sand Packed Column

A. Rajesh Kanna, Sathyanarayana N. Gummadi and G. Suresh Kumar

1 Introduction

Crude oil plays a vital role in our day to day life. The contribution of energy is about 70 % comparatively with other sources of energy. Oil recovery refers to the process by which crude oil is extracted from the cap rock of petroleum reservoir. Crude oil recovery is categorized into three phase namely primary recovery, secondary and tertiary recovery methods. In primary recovery method crude oil is extracted to the surface with help of pressure present in the reservoir. The recovery obtained using the method is about 10–15 %. Once the pressure gets declined no more oil can be recovered to the surface. At this stage secondary recovery method helps to recover oil. In this method, artificially water is pumped at high pressure to build up pressure. Due to pressure difference oil is moved from high pressure to low pressure zone. In water drive method around 20 % of oil is recovered after primary recovery technique (Lazar et al. 2007). Still a considerable amount of oil remains trapped in the cap rocks, which are held due to high interfacial tension and capillary force. Around 50 % of OOIP can be recovered using tertiary recovery method. Injection of different agents like Heat, Polymers, Chemical surfactants and Microbial surfactants to recover crude oil from trapped zone is known as enhanced oil recovery methods. Additionally 20–30 % of oil is recovered after secondary recovery. There are many approaches to improve oil recovery rate. Utilization of microbes and its metabolites is considered to be the most promising methods in oil recovery. Currently numerous studies were carried out using bacteria as EOR

A.R. Kanna · G.S. Kumar
Department of Ocean Engineering, IIT Madras, Chennai, India

S.N. Gummadi (✉)
Department of Biotechnology, Bhupat and Jyoti Mehta School of Biosciences, IIT Madras, Chennai, India
e-mail: gummadi@iitm.ac.in

Prasanna B.D. et al. (eds.), *Biotechnology and Biochemical Engineering*,
DOI 10.1007/978-981-10-1920-3_13

technique. They are well known as microbial enhanced oil recovery (MEOR) where bacteria were used in place of chemical surfactants to retrieve oil. MEOR technique is found to be cost effective among all other EOR methods. In MEOR, microbes and its metabolites help to recover oil by producing bio-surfactant (Li et al. 2002). Bio-surfactants are microbial compounds which are amphiphilic in nature (Fiechter 1992). Bio-surfactants are surface active compounds produced by variety of bacteria (Mukherje et al. 2008). Due to their potential applications in various fields like agriculture, food processing and petroleum industries. Properties of bio-surfactant include surface tension reduction, promoting foaming agents, stable and environment friendly and are effective at high temperature and pH conditions (Makkar and Cameotra 1997; Singh 2012).

In recent years, interests in bio-surfactants have gained steady increase in popularity and have emerged as a promising alternative for chemical surfactants. Bio-surfactants have potential roles in oil industries such as cleaning oil sludge, mobilizing heavy crude oil and managing oil spillage (Desai and Banat 1997). *Pseudomonas aeruginosa*, *Bacillus subtilis* and *Bacillus licheniformis* are some of the well-known bacteria which produce bio-surfactant (Fracchia et al. 2012). Increasing demand for petroleum over recent years and to meet the gap, application of bio-surfactant in oil recovery plays a major role in petroleum industries. However, stability of bio-surfactant at extreme pH and temperature conditions is crucial for its usage in enhanced oil recovery process. Bio-surfactant produced from *P. putida* has the potential to reduce surface and interfacial tension between oil-water mixtures, which is a key factor in oil recovery (Kanna et al. 2014). In our current study, an attempt is made to produce bio-surfactant using *Bacillus subtilis* MTCC 2422. The produced bio-surfactant is further used to recover oil, saturated in sand packed column.

2 Materials and Methods

2.1 *Microorganism*

Bacillus subtilis MTCC 2422 was procured from Microbial Type Culture Collection and Gene Bank (MTCC), Institute of Microbial Technology, India for this current study. The culture was maintained in nutrient agar plates with the following composition (g/L): peptone, 5.0; yeast extract, 2.0; beef extract, 1.0; NaCl, 5.0; agar, 15.0; pH 7.0 ± 0.2, storage temperature −2 to −8 °C.

2.2 *Media and Cultivation Conditions*

Nutrient broth with the following composition (g/L) was used for inoculum preparation. Beef extract, 1.0; yeast extract, 2.0; peptone, 5.0; NaCl, 5.0. *Bacillus subtilis*

(MTCC 2422) grown in Nutrient broth for 8–10 h at 30 °C (A_{600nm} 0.7–0.9) and 2 % (v/v) of the inoculum was used for production of bio-surfactant using mineral salt medium with the following composition (g/L) KNO_3, 0.3; Na_2HPO_4, 0.2; KH_2PO_4, 0.014; NaCl, 0.001; $MgSO_4$, 0.06; $CaCl_2$, 0.004; $FeSO_4$, 0.002; 0.1 mL of trace element solution containing (g/L) $ZnSO_4{\cdot}7H_2O$, 2.32; H_3BO_3, 0.56; $CuSO_4{\cdot}5H_2O$, 1.0; $MnSO_4{\cdot}4H_2O$, 1.78; $Na_2MoO_4{\cdot}2H_2O$, 0.39; $CoCl_2{\cdot}6H_2O$, 0.42; EDTA, 0.5; $NiCl_2{\cdot}6H_2O$, 0.004; KI, 0.66; K_2SO_4, 3.0. Cultivation was performed in 1 L Erlenmeyer flask at 30 °C and 180 rpm in incubator cum shaker. Cell free supernatant was collected after centrifugation at 12,000 × g for 20 min and it was further analyzed for bio-surfactant.

3 Analytical Methods

3.1 *Analysis of Sucrose Concentration*

1 mL of was centrifuged at 12,000 × g for 20 min and the supernatant was collected to determine sucrose concentration using dinitrosalicylic (DNS) method. 1 mL sample was mixed with 25 µl of 3 M HCl and heated at 100 °C for 20 min. 1 mL of DNS reagent was added to the hydrolyzed samples and heated for 100 °C for 10 min. Sucrose concentration was determined by measuring absorbance at 540 nm (Miller 1959).

3.2 *Cell Biomass Determination*

2 mL of sample was subjected to centrifugation for 25 min at 12,325 × g and the supernatant was decanted. The precipitate was washed twice with 0.8 % NaCl and the pellet was dried in hot air oven at 50 °C overnight. The pellet was cooled in desiccator and the dry weight of pellet was measured. The procedure was repeated till concurrent values were obtained.

3.3 *Biosurfactant Determination*

The culture was centrifuged at 12,325 × g to remove bacterial cells. Supernatant was subjected to acid precipitation at pH 2.0 by adding 6 N HCl at 4 °C. The precipitate was pelleted out by centrifugation at 12,325 × g for 25 min, re-suspended in double distilled water and pH was adjusted to 7.0, freeze dried and weighed. The dried surfactant was extracted with dichloromethane and the extract was dried using rotary evaporator under vacuum. The concentrated liquid obtained was the pure form of bio-surfactant.

3.4 Interfacial Tension Analysis

In order to measure interfacial tension, equal volume of crude oil and cell free broth (bio-surfactant) was mixed. IFT was determined by K6 Tensiometer (Kruss GmbH, Hamburg, Germany), using Wilhelmy plate method. 10 mL (5 mL crude oil and 5 mL broth) of sample was placed in the glass container. Measurements were carried out by automatic controller which smoothly pulls down the plate such that it is contacted with the liquid placed. The force acting on the rectangular plate with known length are measured and IFT values were digitally recorded.

3.5 Sand Packed Column Experiment

A sand packed column (a plastic column of 10 cm in length and 2.5 cm diameter) was fabricated to study the oil recovery using biosurfactant produced by the strain. 35 g of fine sand soil was packed in the column and saturated with brine and followed by oil. This mimicks the petroleum reservoir condition. The sand packed column was flooded again with brine until no more oil received at the effluent. 0.5 PV (Pore Volume) of *Bacillus subtilis* MTCC 2422 (OD = 0.23) in mineral salt medium was injected into the column. The column was flooded with water (Secondary flooding) followed by Triton X 100 (Chemical flooding). Produced bio-surfactant was flooded finally to check the recovery rate using various flooding methods. The effluent collected from the outlet of the column gives the amount of oil recovered and determined using standard methods.

3.6 Satistical Analysis

All the experiments were performed at least three times and the values reported are mean of three individual experiments with $p < 0.005$.

4 Results and Discussion

Bacillus subtilis was grown in minimal media with 2 % (w/v) sucrose as the carbon source. Dry cell weight, bio-surfactant production, substrate consumption by the organism were determined every 12 h for a period of 132 h. A time course increase

Fig. 1 Cell dry weight

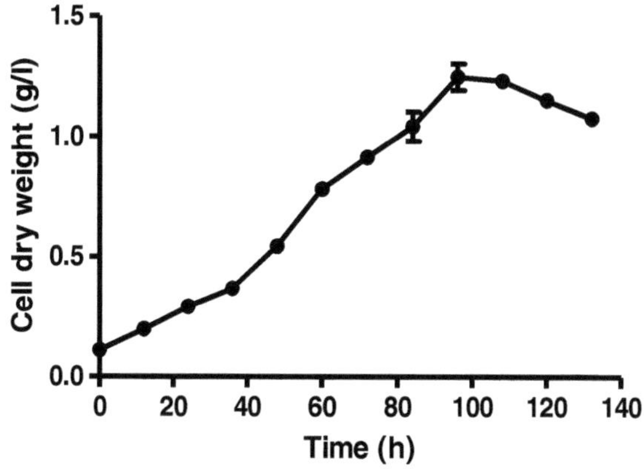

Fig. 2 Bio-surfactant concentration

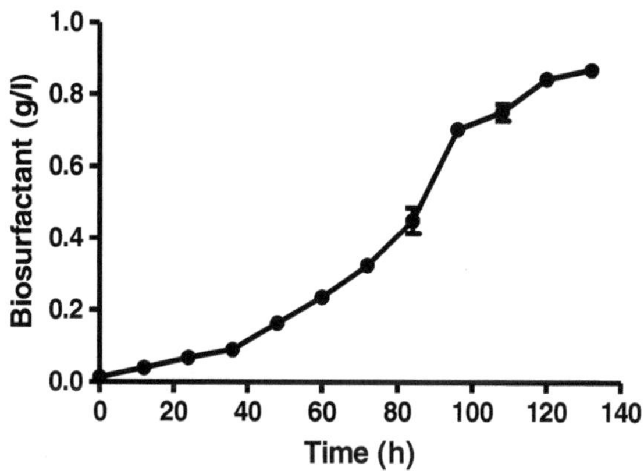

in biosurfact was observed with increase in growth and susbstrate consumption. Maximum biomass obtained was 1.2 g/L at 96 h and maximum bio-surfactant produced at 132 h was 0.84 g/L (Figs. 1 and 2). These results were in agreement with earlier reports such that biomass and bio-surfactant were 0.9 and 1.3 g/L respectively with *Bacillus subtilis* (Priya and Usharani 2009) (Fig. 3).

Maximum production of bio-surfactant was observed in the late log phase. The produced bio-surfactant was further purified by acid precipitation and analyzed for interfacial tension reduction. Interfacial tension (IFT) reduction was directly proportional to the amount of bio-surfactant produced. IFT was initially, 49 mN/m but with increase in fermentation time, the amount of bio-surfactant produced increased gradually resulting in IFT reduction to a final value of 17 mN/m at 132 h (Fig. 4).

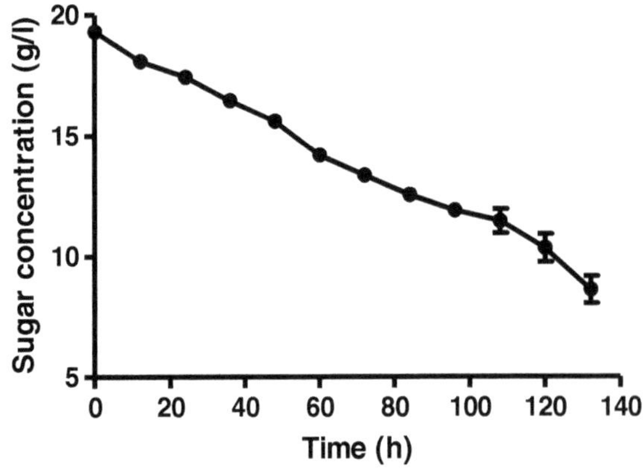

Fig. 3 Reducing sugar concentration

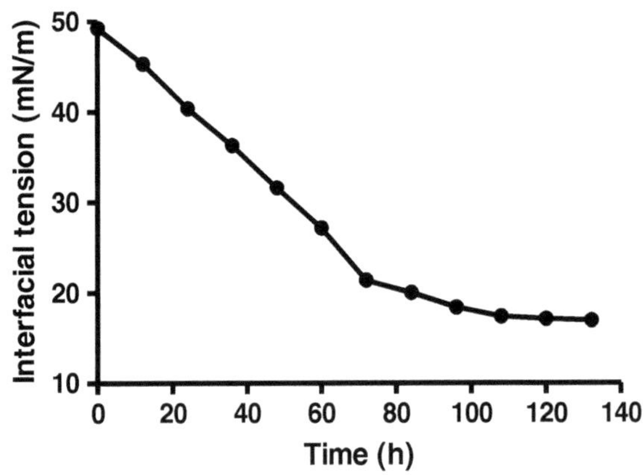

Fig. 4 Interfacial tension profile

Bio-surfactant produced by *Bacillus subtilis* was tested for its possible application microbial enhanced oil recovery by using sand pack column on laboratory scale. Water injection which is secondary recovery method, can only achieve a certain amount of oil recovery. Beyond this, no more oil could be recovered due to high capillary force, which restricts the mobility of oil. As bio-surfactant can reduce the capillary force by reduction of IFT between crude oil and water, we employed this technique to study its influence on oil recovery.

Experimental results showed that water injection initially recovered 27 % of oil and further enhancement of 9 % was observed when bio-surfactant was injected to the column, yielding to a total recovery of 36 % (Fig. 5). Similar results were observed where 38 % of oil was recovered using bio-surfactant production by *Fusarium* sp. (Qazi et al. 2013). Same experiment when repeated with chemical surfactant Triton X 100 gave a total recovery of 34 % (Fig. 6). Since bio-

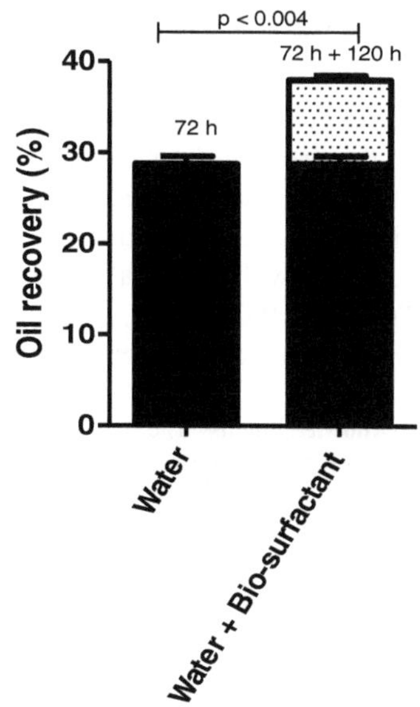

Fig. 5 Oil recovery using bio-surfactant

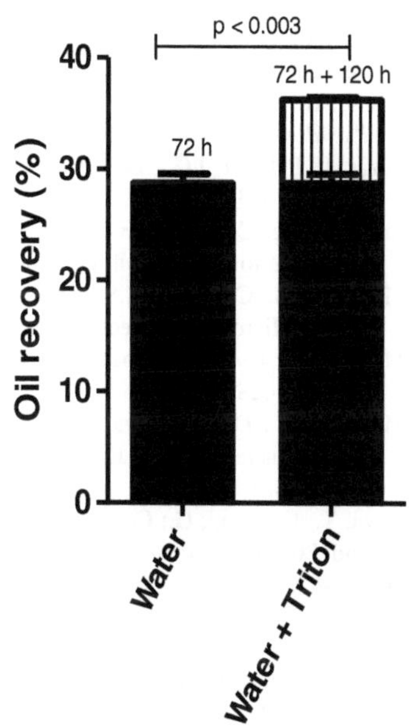

Fig. 6 Oil recovery using Triton

surfactants are less expensive, bio-degradable and able to enhance oil recovery, it can be consider as best alternative for chemical surfactants.

5 Conclusions

Strain *Bacillus subtilis* MTCC 2422 has potential to produce bio-surfactant and IFT of the medium during biosurfact production was reduced from 78 to 32 mN/m. Studies in sand packed column were used to screen and evaluate the possible bio-surfactant application in oil recovery. Around 36 % of oil recovery was achieved with water and bio-surfactant injection similarly 34 % recovery with Triton X 100. Hence, bio-surfactants can be used as promising alternative to chemical surfactants for EOR applications.

Acknowledgments The authors acknowledge Dr. Indumathi M. Nambi, Department of Civil Engineering for interfacial tension analysis. ARK acknowledges MHRD for scholarship.

References

Desai, J.D., Banat, I.M.: Microbial production of surfactants and their commercial Potential. Microbiol. Mol. Biol. Rev. **61**, 47–64 (1997)

Fiechter, A.: Biosurfactants: moving towards industrial application. Trends Biotechnol. **10**, 208–217 (1992)

Fracchia, L., Cavallo, M., Martinooti, M.G., Banat, I.M.: Bio-surfactants and Bio-emulsifiers Biomedical and Related Applications—Present Status and Future Potentials, pp. 325–370. In Tech, Rijeka (2012)

Kanna, R., Gummadi, S.N., Kumar, G.S.: Production and characterization of bio-surfactant by *Pseudomonas putida* MTCC 2467. J. Biol. Sci. **14**(6), 436–445 (2014)

Lazar, I., Petrisor, I.G., Yen, T.F.: Microbial enhanced oil recovery (MEOR). Pet. Sci. Technol. **25**, 1353–1366 (2007)

Li, Q., Kang, C., Wang, H., Liu, C., Zhang, C.: Application of microbial enhanced oil recovery technique to Daqing Oilfield. Biochem. Eng. J. **11**, 197–199 (2002)

Makkar, R.S., Cameotra, S.S.: Bio-surfactant production by a thermophilic *Bacillus subtilis* strain. J. Ind. Microbiol. Biotechnol. **18**, 37–42 (1997)

Miller, G.L.: Use of dinitrosalicylic acid reagent for determination of reducing sugar. Anal. Chem. **31**, 426–428 (1959)

Mukherje, S., Das, P., Sivapathasekaran, C., Sen, R.: Enhanced production of bio-surfactant by marine bacterium on statistical screening of nutritional parameters. J. Biochem. Eng. **42**, 254–260 (2008)

Priya, T., Usharani, G.: Comparative study for biosurfactant production by using *Bacillus subtilis* and *Pseudomonas aeruginosa*. Bot. Res. **2**, 284–287 (2009)

Qazi, M.A., Malik, Z.A., Qureshi, G.D., Hameed, A., Ahmed, S.: Yeast extract as the most preferable substrate for optimized biosurfactant production by *rhl*B gene positive *Pseudomonas putida* SOL-10 isolate. J. Bioremed. Biodeg. **4**, 1–10 (2013)

Singh, V.: Biosurfactant—isolation, production, purification & significance. J. Sci. Ind. Res. **2**, 1–4 (2012)

Extraction of Polyphenols from Orange Peel by Solvent Extraction and Microbial Assisted Extraction and Comparison of Extraction Efficiency

Prabha Hegde, Pushpa Agrawal and Praveen Kumar Gupta

1 Introduction

One of major secondary metabolites of the plants are Polyphenols, which contains more than one phenol unit as building block of molecule. Fruits, vegetables, leguminous plants and some cereals are rich sources of polyphenols. Polyphenols are often present in higher concentration in the outer non edible part of the fruits like peel compared to inner edible part (Gianmaria 2011; Vinson and Hontz 1995; Wolfe 2003; Vinson 1998, 2001). Orange peels contains high concentration of polyphenols which makes orange as high antioxidant source, (Hegazy and Ibrahium 2012). Orange peels are rich in polyphenols. Polyphenols are abundantly present in our diet (Scalbert and Williamson 2000). The benefits of dietary polyphenols have been studied extensively over the last decade. Polyphenols have been established to play a significant role in preventing diseases like cardiovascular diseases, cancer and other diseases (Duda-Chodak and Tarko 2007).

In this research work polyphenols are extracted by using solvent and microbial assisted extraction. Total polyphenols was estimated by using Folin Ciocalteu spectrophotometer method. The major polyphenols were estimated individually by using High Pressure Liquid Chromatography (HPLC) technique.

P. Hegde (✉) · P. Agrawal · P.K. Gupta
Department of Biotechnology, R V College of Engineering, Mysore Road, RV Vidyaniketan Post, Bangalore 560059, India
e-mail: prabha_hegde@yahoo.com

Prasanna B.D. et al. (eds.), *Biotechnology and Biochemical Engineering*,
DOI 10.1007/978-981-10-1920-3_14

2 Materials and Methods

2.1 Materials

Gallic acid monohydrate, Caffeic acid, p-Coumaric acid (+)-Catechin hydrate, Epicatechin were purchased from Sigma–Aldrich, Methanol (HPLC) Sodium carbonate and Phosphoric acid were obtained from Merck. Folin Ciocalteu reagent, Soybean Casein Digest Broth, Potato Dextrose Broth were obtained from HiMedia. *Staphylococcus aureus* (ATCC 6538), *E. coli* (ATCC 8739) and *Aspergillus niger* (ATCC9642) was procured from National Chemical Laboratory, Pune, India.

2.2 Samples

Three types of oranges namely Coorg, Nagpur and Kinnow were used for the study. Outer Peels of fresh oranges were cut into small pieces of $\sim$2 mm × 2 mm dimension. 50 g of the fresh peel was dried by each of the following different techniques, viz., freeze dried in lyophilizer at −50 °C for 36 h (Li et al. 2006a, b), oven dried at 50 °C for 36 h and sundried for 2 days in bright sunlight. Abbreviation of the sample used are tabulated in Table 1.

2.3 Extraction Method

Dried samples were chemically extracted using a mixture of water and methanol (1:1) (Li et al. 2006a, b), at 95 °C for 3 h (Hegde et al. 2015). After extraction samples were filtered using Whatman filter paper number 1, and filtrate are stored in the refrigerator.

Microbiological extraction was performed by using sterilised Soybean Casein Digest Broth (for bacteria) and Potato Dextrose Broth (for fungi). Different organism like *Staphylococcus aureus*, *E. coli* and *Aspergillus niger* were used (Arora and Kaur 2013; Armando et al. 2008). Optimum results were obtained using

Table 1 Name of the samples

Variety	Nagpur	Nagpur	Nagpur	Coorg	Coorg	Coorg	Kinnow	Kinnow	Kinnow
Drying technique	Oven	Sun	Freeze	Oven	Sun	Freeze	Oven	Sun	Freeze
Abbreviation	NO	NS	NF	CO	CS	CF	KO	KS	KF
Chemical extraction	NOC	NSC	NFC	COC	CSC	CFC	KOC	KSC	KFC
Micro extraction	NOM	NSM	NFM	COM	CSM	CFM	KOM	KSM	KFM

Potato Dextrose Broth, *Aspergillus niger* organism, incubation temperature of 20 °C for 48 h using 2 g of sample in 100 mL of the media. After 48 h samples were filtered using 0.2 micron nylon 47 mm disc membranes under vacuum and stored in the refrigerator.

2.4 Estimation of Total Polyphenols by Spectrophotometer Method

Total Polyphenol content was estimated by modified Folin Ciocalteu method. In this test, two strong inorganic oxidants namely phosphotungstic and phosphomolybdic acids (Stevanato 2004) reduce the molecules by chemical oxidation and then Folin Ciocalteu develops the color which is read at 765 nm in spectrophotometer. Gallic acid is used as standard and a standard curve was obtained using 10–100 ppm concentration solutions with a coefficient of correlation of 0.99 (Fig. 1). 1 mL of known dilution of the sample that were previously made up with methanol were taken in 10 mL volumetric flasks. 0.5 mL of Folin Ciocalteu reagent was added and allowed to stand for 3 min. 1.5 mL of 10 % Sodium carbonate was added and volume made up to 10 mL with water. The solutions were heated at 50 °C for 16 min. Samples were cooled and absorption was measured at 765 nm using a UV visible spectrophotometer (Perkin Elmer, Lamda35) with water as blank. Each sample was analyzed in triplicate. Sample absorbance was compared with standard graph obtained for Gallic acid from which concentration of total polyphenols was estimated.

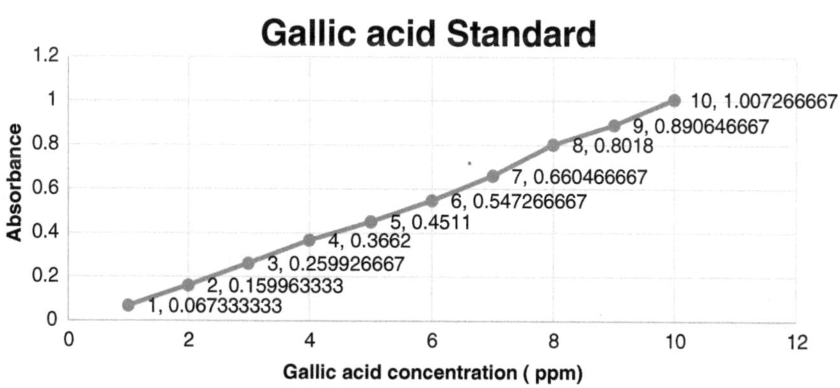

Fig. 1 Gallic acid standard graph (n = 3). Gallic acid standard solution were prepared from 1 to 10 ppm

2.5 Estimation of Individual Polyphenols Using HPLC Method

Polyphenol standards like Gallic acid, Catechin, Epicatechin, Caffeic acid and p-Coumaric acid were used for characterization of the phenolics in the extraction. A linear curve was produced for each of these standards and the linear calibration curve within the concentration studied had a coefficient of correlation ranging from 0.998 to 0.999.

Analysis was performed using phosphoric acid (0.1 %) as mobile phase A and 100 % methanol (HPLC) as solvent B. The elution conditions were: 0–30 min from 5 % B to 80 % B; 30–33 min 80 % B; 33–35 min from 80 % B to 5 % B; flow rate = 0.8 mL/min. The operating conditions were: column temperature, 40 °C; injection volume of 10 μL and a C18 waters Symmetry column was used. The detection wavelength was 280 nm for Gallic acid. Catechin and Epicatechin, 320 nm for Caffeic acid and p-Coumaric acid (Šeruga et al. 2011; Antoanela et al. 2011)

Extracted samples were removed from refrigeration and allow to come to room temperature, 1 mL of the sample was diluted to 10 mL using the aqueous methanol (1:1) as diluent, this diluted samples were filtered through Puradisc 25 mm syringe filters from Whatman before injecting into HPLC. Each samples were injected three times.

3 Results and Discussion

As Gerhard (EuroFood Chem, "Food for Future" in 2009) states, citrus peels shows higher concentration of polyphenols compared to seeds. So only peels were considered for this study. Solvent extraction was performed by using aqueous methanol as solvent (Li et al. 2006a, b) with optimized conditions (Hegde et al. 2015). Armando et al. used *Aspergillus niger* to extract Ellagic acid, a polyphenol from pomegranate residues, in this study *Aspergillus niger* was used to extract polyphenols from orange peel.

Solvent extraction and the Microbial assisted extraction produced different results. Different varieties of oranges and their drying technique also had an effect on content of total polyphenols content and the individual identified polyphenols in the orange peels. Total Polyphenols content in orange peel samples extracted by both solvent and microbial assisted extraction method were estimated by modified Folin Ciocalteu test method and calculated equivalent to Gallic acid. Results are tabulated in Fig. 2.

Several thousand molecules having a polyphenols structure have been identified in higher plants, and several hundred are found in edible plants (Manach et al. 2004). It is difficult identify and quantify each individual polyphenols so HPLC test method is used to identify and quantify few important polyphenols like Gallic acid, Catechin, Epicatechin, Caffeic acid and p-Coumaric. Five standards from Sigma

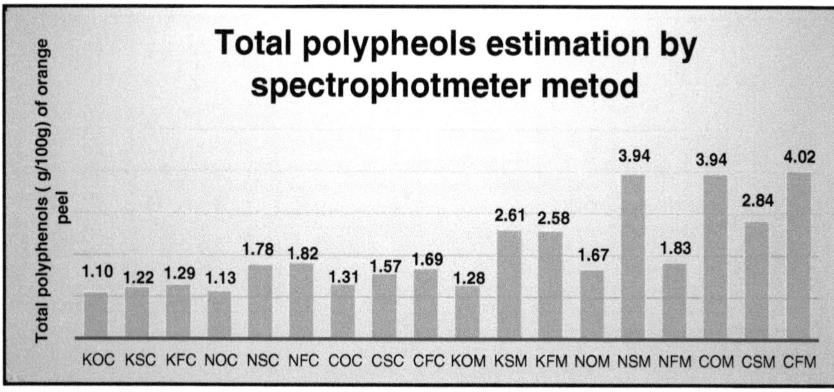

Fig. 2 Total polyphenol content of orange peel (n = 9). Orange peel dried under different conditions were extracted by chemical and micro extraction method

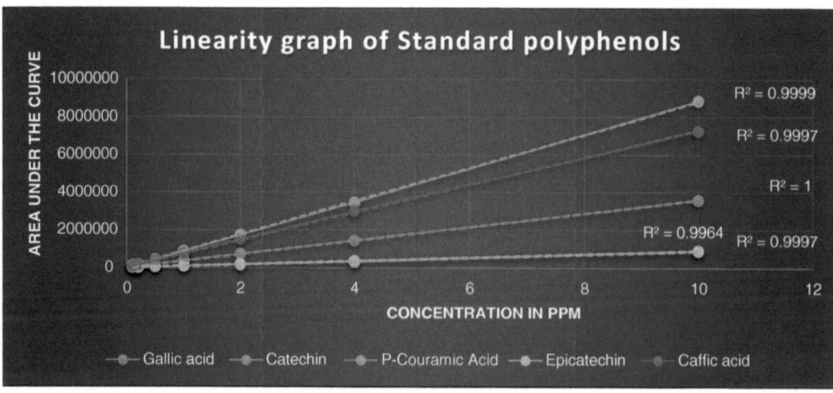

Fig. 3 Gallic acid, Catechin, Epicatechin P Couramic acid and Caffic acid standard graph (n = 5). These standard solution were prepared from 0.1 to 10 ppm and analysed by HPLC. All the standard has a R2 of minimum 0.9996

Aldrich were used to create a standard curve for each of these standards and Fig. 3 shows the linearity of these standards

Results very clearly demonstrates that total polyphenols content are more in the microbial assisted extraction compared to chemical extraction. Results from the chemical extraction shows that total polyphenols in Nagpur freeze dried shows maximum followed by Nagpur sundried. Coorg orange peel showed higher concentration of total polyphenols and individuals polyphenols compared to Nagpur and Kinnow. Different drying had a very less effect on the total polyphenols content, though sundried and freeze dried showed a slightly higher value. In the microbiological extraction also Coorg orange had a highest concentration. Out of the five individual polyphenols quantified Catechin showed highest concentration in all the samples (Fig. 2).

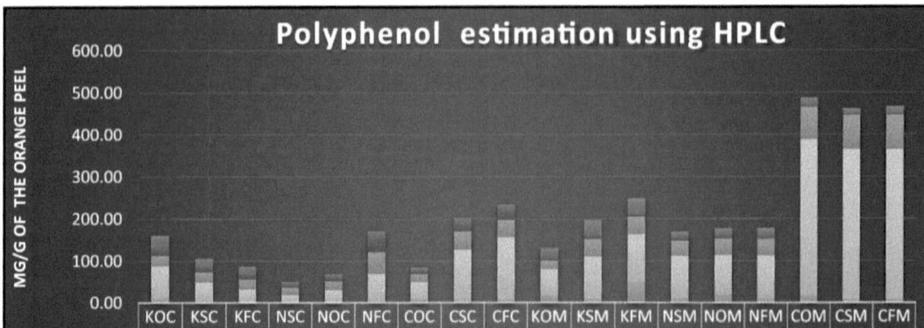

	KOC	KSC	KFC	NSC	NOC	NFC	COC	CSC	CFC	KOM	KSM	KFM	NSM	NOM	NFM	COM	CSM	CFM
P couramic	23.09	15.18	15.06	5.01	5.71	6.18	5.67	6.89	6.28	22.15	33.06	34.03	11.15	11.54	13.98	14.54	9.93	14.74
Caffic acid	25.05	16.35	14.60	8.20	10.70	43.60	8.45	24.50	29.71	8.01	12.03	9.21	11.00	12.54	12.48	7.46	5.52	5.91
Epicatechin	23.99	23.74	24.25	18.50	21.66	51.51	19.32	42.99	40.41	20.33	41.56	41.06	35.10	39.27	38.58	77.43	81.38	81.38
Catechin	78.28	46.02	30.64	16.32	30.04	65.34	43.67	122.9	149.5	61.20	100.2	113.2	100.7	91.74	100.7	369.9	356.1	356.1
Gallic acid	10.28	4.01	2.15	2.33	1.11	4.19	7.03	4.07	7.14	19.17	10.24	50.38	10.99	21.49	11.73	18.47	8.92	8.92

Fig. 4 Gallic acid, Catechin, Epicatechin, P Couramic acid and Caffic acid contents of the orange peels by chemical extraction and microbiological extraction (n = 5)

4 Conclusion

Maria et al. (2013) showed that citrus peels exhibit quite high antioxidant activity, which together with polyphenols contained makes them a valuable source for natural antioxidants and citrus flavonoids in the cosmetic and food industries. As per study by the Gerhard (EuroFood Chem, "Food for Future" in 2009) polyphenols have remarkable antioxidant and radical scavenging properties, so it is worth considering polyphenols as additive for food products and dietary supplement. As per Market research (study Transparency Market Research. *April 01, 2015)* polyphenol consumption can increase up to 21,033 tons by the end of 2018 with CGAR of 8 %. But there are challenges as there is increased pressure on land for agriculture, and research is focused on finding out the exact class of plant for production. Our work is showing a way forward for this problem as the source for polyphenol extraction is orange waste, and if microbiological extraction considered, it is ecofriendly also as there is no use of solvent and heat involved in this extraction.

Though total polyphenols content of the Coorg oranges ranged from 2.84 to 4.02, Kinnow ranged from 1.28 to 2.68 and Nagpur had 1.67 to 3.94. There could be geographical and seasonal effect are natural variations which may affect the polyphenols concentration in the orange peels (Figs. 3 and 4).

References

Antoanela, P., Negreanu-Pirjol, T., Rosca, C., Arcus, M., Bucur, L., Istudor, V.: HPLC analysis of polyphenols and antioxidant capacity determination of *Scirpus holoschoenus* L. rhizome. Ovidius Univdersity Ann. Chem. **22**(1), 62–66 (2011)

Armando, R., Aguilera-Carbo, A., Rodriguez, R., Martinez, J.L., Garza, Y., Aguilar, C.N.: Ellagic acid production by *Aspergillus niger* in solid state fermentation of pomegranate residues. J. Ind. Microbiol. Biotechnol. **35**, 507–513 (2008)

Arora, M., Kaur, P.: Antimicrobial and antioxidant activity of orange pulp and peel. Int. J. Sci. Res. (IJSR). **2**(11), 2319–7064 (2013). ISSN (online)

Duda-Chodak, A., Tarko, T.: Antioxidant properties of different fruits seeds and peels. Acta Sci. Pol. Technol. Aliment. **6**(3), 29–36 (2007)

Gerhard, K.: Antioxidant activity and total polyphenols in citrus fruit peels and seeds. In: Proceeding of Food for future- the Contribution of Chemistry to Improvement of Food Quality (EuroFoodChemXV). Copenhagen, Denmark, 5–8 July 2009

Gianmaria, F.F., Amato, I., Ingenito, A., Zarrelli, A., Pinto, G., Pollio, A.: Plant polyphenols and their anti carcinogenic properties: a review. Molecules **16**, 1486–1507 (2011)

Hegazy, A.E., Ibrahium, M.I.: Antioxidant activities of orange peel extracts. World Appl. Sci. J. **18**(5), 684–688 (2012)

Hegde, P., Agrawal, P., Gupta, P.K.: Isolation and optimization of polyphenols from the peels of orange fruit. J. Chem. Pharm. Sci. **8**(3) (2015)

Li, B.B., Smith, B., Hossain, M.M.: Extraction of phenolics from citrus fruits peels II enzyme assisted extraction method. Sep. Purif. Technol. **48**, 189–196 (2006a)

Li, B.B., Smith, B., Hossain, M.M.: Extraction of phenolics from citrus peels, I solvent extraction method. Sep. Purif. Technol. **48**, 182–188 (2006b)

Manach, C., Scalbert, A., Morand, C., Rémésy, C., Jiménez, L.: Polyphenols: food sources and bioavailability. Am. Soc. Clin. Nutr. **7**(9), 727–747 (2004)

Maria, K., Kirova, E., Alexandrova, S.: Natural antioxidants from Citrus Mandarin peels, extraction of polyphenols; effect of operational conditions on total polyphenols contents and antioxidant activity. J. Chem. Technol. Metall. **48**(1), 35–41 (2013)

Scalbert, A., Williamson, G.: Dietary intake & bioavailability of polyphenols. J. Nutr. **130**, 2073S–2085S (2000)

Šeruga, M., Novak, I., Jakobek, L.: Determination of polyphenols content and antioxidant activity of some red wines by differential pulse voltammetry. HPLC Spectrophotometric Method Food Chem. **124**, 1208–1216 (2011)

Stevanato, R., Fabris, S., Momo, F.: New enzymatic method for the determination of total phenolic content in tea and wine. J. Agric. Food Chem. **52**, 6287–6293 (2004)

Transparency Market Research: Polyphenols Market is anticipated to reach a total consumption of 21,032.7 tons by the end of 2018, will demonstrate a CAGR of 6.1 % during the forecast period of 2012 to 2018: Transparency Market Research. *April 01, 2015 09:36 ET*https://globenewswire.com/news-release/2015/04/01/721169/10127269/en/Polyphenols-Market-is-anticipated-to-reach-a-total-consumption-of-21-032-7-tons-by-the-end-of-2018-will-demonstrate-a-CAGR-of-6-1-during-the-forecast-period-of-2012-to-2018-Transpa.html. Cited on 21 Feb 2016

Vinson, J.A., Hontz, B.A.: Phenol antioxidant index: comparative antioxidant effectiveness of red and white wines. J. Agric. Food Chem. **43**, 401–403 (1995)

Vinson, J.A., Hao, Y., Su, X., Zubik, L.: Phenol antioxidant quantity and quality in food: vegetables. J. Agric. Food Chem. **46**, 3630–3634 (1998)

Vinson, J.A., Su, X., Zubik, L., Bose, P.: Phenol antioxidant quantity and quality in food: fruits. J. Agric. Food Chem. **49**, 5315–5321 (2001)

Wolfe, K., Wu, X., Liu, R.H.: Antioxidant activity of apple peels. J. Agric. Food Chem. **51**, 609–614 (2003)

Enzymatic Concentration of *n*−3 Polyunsaturated Fatty Acids from Indian Sardine Oil

Charanyaa Sampath, N. Anita, B.D. Prasanna and Iyyaswami Regupathi

1 Introduction

The demand for healthy food and nutrition are becoming a need for everyone to build up the immunity for different types of diseases. Therefore, there is a search for the production of compounds which prevent these diseases. Many reports have studied the beneficial properties of *n*−3 PUFA namely Eicosapentaenoic acid (EPA) and Docosahexaenoic acid (DHA). Researches over the last decades have shown positive effects of fish oils on the cognitive development and vision enhancement in the newborns (Colombo et al. 2004). The long chain fatty acids like EPA and DHA have been shown to reduce certain types of cancers, diabetes, mental health disorders and asthma (Alasalvar et al. 2002; Nettleton and Katz 2005). The presence of large quantities of these two long chain polyunsaturated fatty acids (PUFA) in Sardine oil (Pike and Jackson 2010) has induced immense interest in researchers to look for methods to concentrate *n*−3 PUFA from sardine oil. In response to this demand, pharmaceutical industries used various techniques, namely, chromatographic separation, fractional distillation, low temperature crystallization, supercritical fluid extraction, and urea complexation. However, most of these methods produce PUFA in the form of free fatty acids (FFA) or alkyl esters which are nutritionally unfavorable and are known to occur at high temperatures. Since *n*−3 PUFA are sensitive to heat and oxidation, a hunt for mild conditions which leads to the concentration of EPA and DHA have become essential. Several papers describing the use of lipase for hydrolyzing the glycerides in fish oil are available. Wanasundara and Shahidi (1998) have reported the use of *Candida cylindracea* lipase for the enrichment of *n*−3 PUFA in seal blubber oil. The *n*−3

C. Sampath (✉) · N. Anita · B.D. Prasanna · I. Regupathi
Department of Chemical Engineering, NITK, Mangaluru, Karnataka, India
e-mail: charansampath.2853@gmail.com

Prasanna B.D. et al. (eds.), *Biotechnology and Biochemical Engineering*,
DOI 10.1007/978-981-10-1920-3_15

PUFA content increased from 28.0 % (w/v) to 43.5 % after 70 h enzymatic treatment. Hoshino et al. (1990) have claimed twofold increase in $n-3$ PUFA after *Candida cylindracea* lipase treatment in the case of refined Sardine oil. Shimada et al. (1994) have reported that *Geotrichum candidum* lipase increased the $n-3$ PUFA content to 48.7 % in tuna oil. However, no literature has studied the hydrolysis of Indian Sardine oil. In this study, maximum activity of *Candida rugosa* lipase (CRL) was found by optimizing the various operation parameters. The Indian Sardine fish oil was subjected to hydrolysis by this CRL to concentrate the $n-3$ PUFA content.

2 Materials and Methods

2.1 *Materials*

Crude Indian sardine oil was obtained from Mukka fish oil industries, Mukka, which was obtained by a conventional pressing method. The refined oil was achieved through degumming, neutralization, bleaching and deodorization according to Cmolik and Pokorny (2000). Refined oil was stored under nitrogen at −20 °C in dark container until use. CRL was obtained from Sigma Aldrich, India. All other chemicals and solvents were of reagent grade and were used without further purification.

2.2 *Methods*

The optimal activity of the CRL was obtained by studying the parameters like pH, temperature, water- oil ratio, solvent- oil ratio and time. The reaction mixture containing 1 g of oil and 0.325 mg/mL of CRL was subjected to pH ranging from (5.5, 6, 6.5, 7, 7.5, 8), temperature (25, 30, 35, 40, 45 °C), oil to water ratio (1:1, 1:4, 1:8, 1:10, 1:12), oil to solvent ratio (1:0.5, 1:1, 1:1.5, 1:2) and time (10, 20, 30, 40, 50, 60 min) by shaking at 500 rpm at room temperature. Enzymatic activity was obtained by the determination of the liberated FFA through titration with 0.1 N KOH using phenolphthalein as the end point indicator. The activity of CRL was calculated according to the official method of American Oil Chemists' Society (AOCS) (2009) methodologies, (Cd 3d-63). The hydrolysis was performed under the optimized reaction conditions of CRL for 30 min and the reaction was stopped by addition of 10 mL ethanol. The hydrolyzed oil was subjected to solvent extraction using methanol in the ratio 1:1 oil to methanol (w/w) in order to remove the FFA released during the hydrolysis reaction. The solvent extracted oil was studied for its iodine value (AOCS (Cd 1c-85) 2009) which was compared with the iodine value of the refined oil measured before hydrolysis.

3 Results and Discussion

Stability of the lipase was checked at different pH, temperatures, time, water concentration and solvent amount. This study is important since the information it reveals about the optimum processing conditions is for the efficient hydrolysis of sardine oil.

3.1 *Effect of pH on Activity of CRL*

The activity of CRL was studied at room temperature by dissolving it in the phosphate buffer ranging in pH from 5.5 to 8 as given in Fig. 1.

As seen in the Fig. 1, lipase showed the maximum activity of 90.9 mg KOH/g of oil at pH 7. This is because they generally contain residues of amino acids at the active site which bind efficiently to the substrate at neutral pH. CRL was unstable in alkaline and extremely acidic pH because of the structural changes in the proteins due to the pH variation. In alkaline solutions (pH > 8.0), the partial damage of cysteine residues caused by β-elimination could result in the reduced activity of CRL and in acidic solutions (pH < 4.0) the labile peptide bonds sometimes found next to aspartic acid residues may be hydrolyzed (Akova and Ustun 2000). Besides, at acidic pH, the enzyme may form high molecular weight aggregates due to the unspecific associations with other proteins because of its hydrophobic nature (Montero et al. 1993). The result of this study was similar with that of Montero et al. (1993), where the soluble CRL was active between pH 6.2 and 7.7. Another study by Fadologlu and Soylemez (1997) stated the optimum pH for the soluble CRL as 7.0.

3.2 *Effect of Temperature on the Activity of CRL*

Tests to decide the effect of temperature on lipase activity were done at a temperature range of 25–45 °C, and it was taken in that CRL showed a maximum

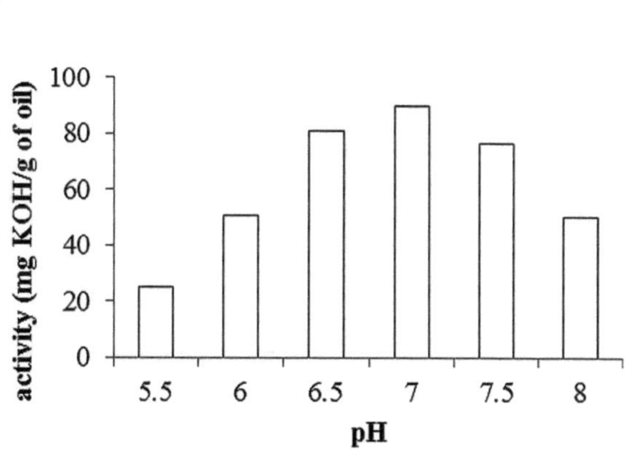

Fig. 1 Effect of pH of buffer on activity of CRL at room temperature

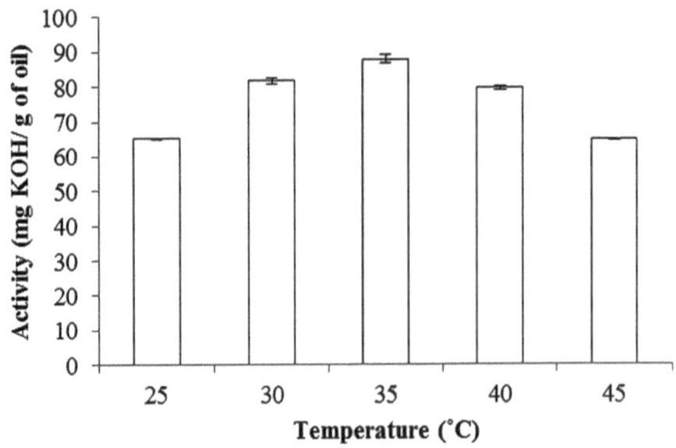

Fig. 2 Effect of temperature on the activity of CRL at pH 7

activity of 88.7 mg KOH/g of oil at a temperature of 35 °C in pH 7.0 phosphate buffer. However, 9.4 and 26.35 % of the activity was lost at 40 °C and 45 °C, respectively (Fig. 2). This is due to the denaturation of the enzyme structure at these temperatures. Several researchers have reported the optimum temperature of soluble CRL at 37 °C (Montero et al. 1993; Xu et al. 1995). Also, Montero et al. (1993) found that treatment at higher temperatures led to the inactivation of the enzyme. According to Fadologlu (1996), CRL exhibits its maximal activity at 40 °C. In this study, enzyme showed considerable activity in a temperature of 35 °C.

3.3 *Effect of Oil to Water Ratio*

CRL catalyse the splitting of the ester bonds of triglycerides with the consumption of water molecules. Hence it gets significant to consider the quantity of water required for hydrolysis as it increases the functionality of the enzyme at the interphase in the biphasic solvent system. From Fig. 3, it is clearly seen that the hydrolytic activity was highest (88.5 mg KOH/g of oil) at the oil to water ratio of 1:10 (w/v). The ratios less than 1:10 (w/v) and above 1:10 (w/v) showed lesser activities which could be because the lesser quantity of water leads to the reduced

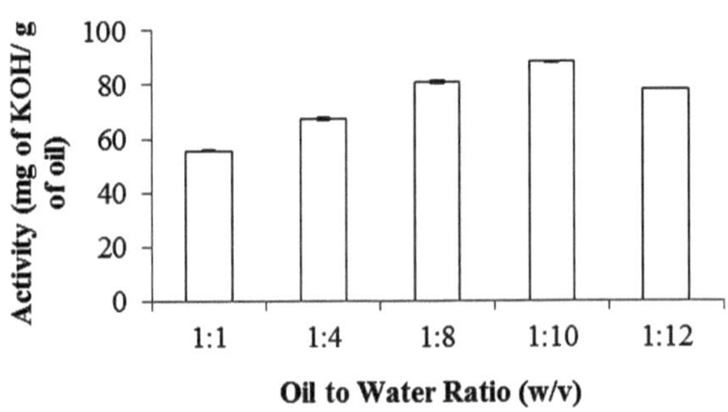

Fig. 3 Effect of oil to water ratio (w/v) on the activity of CRL at pH 7, temperature 35 °C

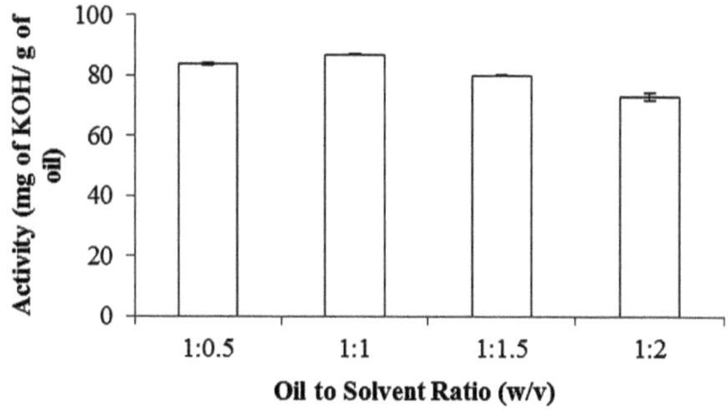

Fig. 4 Effect of oil to solvent ratio (w/v) on the activity of CRL at pH 7, temperature 35 °C

rate of conversion, whereas higher quantity of water leads to a thicker water layer around the CRL which increases the flexibility of CRL to interact with the solvents,causing denaturation. (Aditi et al. 2014). Han and Rhee (1986) stated similar reaction conditions for the hydrolysis of olive oil catalyzed by CRL enzyme in the range of pH 6.5–7.1, temperature 30–35 °C and 0.72–9.78 (v/v) oil to water ratio.

3.4 *Effect of Oil to Solvent Ratio*

The enzyme displayed a maximum activity of 86.7 mg KOH/g of oil at 1:1 (w/v) oil to solvent ratio, as is shown in the Fig. 4. The isooctane due to its non polarity, acts as an organic phase and completely dissolves the non polar triglycerides formed during hydrolysis and separates them with great ease (Yadav and Devi 2004). The enzyme activity is found to decrease beyond 1:1 (w/v) oil to solvent ratio due to the excess amount of iso octane which reduces the availability of the active site of enzyme to the oil. (Aditi et al. 2014). The presence of iso octane results in an increased interfacial area of the oil water system due to the reduced effect of viscosity of the oil.(Aditi et al. 2014). The non polarity of iso octane reduce the enzyme inhibition without stripping off the essential water from CRL during hydrolysis and helps in maintaining a monolayer of water around the CRL (Klibanov 1989).

3.5 *Effect of Time*

Figure 5 shows the time course of hydrolysis where the CRL was shown to give maximum product formation at 30 min equivalent to 85.4 mg of KOH/g of oil. It is clear from the trend observed in Fig. 5, that with the initial increase in time the product formation was increased up to 30 min and a further increase in time up to 1.5 h did not lead to any improvement in the product formation. The progress of biocatalytic reactions is almost never linear and is given to slow down with time

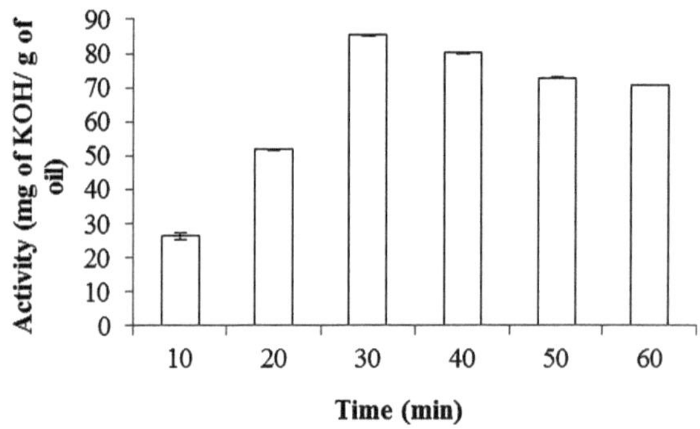

Fig. 5 Effect of variation in time (minutes) on the activity of CRL at pH 7, temperature 35 °C

due to the decrease in oil concentration, the increase in product concentration or the inactivation of the biocatalyst (Gardossi et al. 2010).

Thus, the most suitable reaction conditions for hydrolysis of sardine fish oil with CRL in solvent system were obtained at pH 7.0, temperature 35 °C, and oil to water ratio of 1:10 (w/v), oil to solvent ratio of 1:1 (w/v), and time 30 min. The activity of lipase for optimized parameters was found to be 85.4 for CRL in μmoles of FFAs/ml.

3.6 Enhancement of n−3 *PUFA in Oil*

The iodine value of oil before hydrolysis was found to be 145.45 and after hydrolysis it was 162.29. This significant increase in iodine value number indicates the enhancement of unsaturated fatty acids in the oil. The increase in the unsaturation is probably due to the enhancement of nutritionally important *n*−3 PUFA like EPA and DHA. This is because CRL has fatty acid chain length specificity showing an increased discrimination against the long chain PUFA like C18–C22. CRL hydrolyses the short chain fatty acids, saturated fatty acids and mono unsaturated fatty acids because of the reduced steric hinderances when linked to the glycerol backbone acids leading to the protection and concentration of EPA and DHA (Okada and Morrissey 2006).

4 Conclusions

In this study, the optimal activity of CRL was found after which the hydrolysis of oil with CRL in the optimized conditions was carried out followed by FFA removal which resulted in an enhancement of *n*−3 PUFA in Indian sardine oil. It should be noted that the hydrolysis of oil for 30 min resulted in a significant increase in *n*−3 PUFA after which there was not much increase observed. For commercial purposes, 30 min might be an optimum reaction time for producing *n*−3 PUFA concentrates.

Hence lipase catalysed hydrolysis was considered to be a beneficial and feasible method for the enhancement of $n-3$ PUFAs from Indian sardine oil for the use in nutraceuticals and other products.

References

Aditi, S., Chaurasia, S.P., Dalai, A.K.: Non-selective hydrolysis of tuna fish oil for producing free fatty acids containing docosahexaenoic acid. Can. J. Chem. Eng. **92**, 344–354 (2014)

Akova, A., Ustun, G.: Activity and adsorption of lipase from *Nigella sativa* seeds on celite at different pH values. Biotechnol. Lett. **22**, 355–359 (2000)

Alasalvar, C., Shahidi, F., Quantick, P.: Food and health applications of marine nutraceuticals: a review. Seafoods Qual. Technol. Nutraceutical Appl. 175–195 (2002)

AOCS: Official methods and recommended practices of the American oil chemists. Soc. Sampling Anal. Fats Oils (2009)

Cmolik, J., Pokorny, J.: Physical refining of edible oils. Eur. J. Lipid Sci. Technol. **102**, 472–486 (2000)

Colombo, J., Shaddy, D.J., Anderson, C.J., Blaga, O.M., Kannass, K.N., Kundurthi, S.: Maternal DHA and the development of attention in infancy and toddlerhood. Child Dev. **75**(4), 1254–1267 (2004)

Fadologlu, S.: Kinetics of olive oil hydrolysis by free and immobilised *Candida rugosa* lipase. Ph. D thesis, University of Gaziantep (1996)

Fadologlu, S., Soylemez, Z.: Kinetics of lipase catalyzed hydrolysis of olive oil. Food Res. Int. **30**, 171–175 (1997)

Gardossi, L., Poulsen, P.B., Ballesteros, A., Hult, K., Svedas, V.K., Vasic-Racki, D., Carrea, G., Magnusson, A., Schmid, A., Wohlgemuth, R., Halling, P.J.: Guidelines for reporting of biocatalytic reactions. Trends Biotechnol. **28**, 171–180 (2010)

Han, D., Rhee, J.S.: Lipase-catalyzed hydrolysis of milk fat in lecithin reverse micelles. Biotechnol. Bioeng. **28**, 1250–1255 (1986)

Hoshino, T., Yamane, T., Shimazu, S.: Selective hydrolysis of fish oil by lipase to concentrate $n-3$ polyunsaturated fatty acids. Agric. Biol. Chem. **54**, 1459–1467 (1990)

Klibanov, M.A: Enzymatic catalysis in anhydrous organic solvents. Trends Biochem Sci. **14**(4), 141–144 (1989)

Montero, S., Blanco, A., Virto, M., Ladenta, L.C., Agud, I., Solozabal, R., Lascaray, J.M., Renobales, M., Llama, M.J., Serra, J.L.: Immobilization of *Candida rugosa* lipase and some properties of the immobilized enzyme. Enzyme Microb. Technol. **15**, 239–247 (1993)

Nettleton, J.A., Katz, R.: $n-3$ long-chain polyunsaturated fatty acids in type 2 diabetes: a review. J. Am. Diet. Assoc. **105**(3), 428–440 (2005)

Okada, T., Morrissey, M.T.: Production of $n-3$ polyunsaturated fatty acid concentrate from sardine oil by lipase-catalyzed hydrolysis. Food Chem. **103**, 1411–1419 (2006)

Pike, I.H., Jackson, A.: Fish oil: production and use now and in the future. Lipid Technol. **22**(3), 59–61 (2010)

Shimada, Y., et al.: Enrichment of polyunsaturated fatty acids with *Geotrichum candidum* lipase. J. Am. Oil Chemist's Soc. **71**(9), 951–954 (1994)

Wanasundara, U.N., Shahidi, F.: Lipase-assisted concentration of $n-3$ polyunsaturated fatty acids in acylglycerols from marine oils. J. Am. Oil Chemist's Soc. **75**(8), 945–951 (1998)

Xu, H., Li, M., He, B.: Immobilization of *Candida cylindracea* lipase on methyl acrylate-divinyl benzene copolymer and its derivatives. Enzyme Microb. Technol. **17**, 194–199 (1995)

Yadav, G.D., Devi, K.M.: Kinetics of hydrolysis of tetrahydrofurfuryl butyrate in a three phase system containing immobilized lipase from *Candida antarctica*. J. Biochem. Eng. **17**, 57–63 (2004)

High-Throughput Screening of Cell Repellent Substrate Chemistry for Application in Expanded Bed Adsorption Chromatography

Vikas Yelemane, Martin Kangwa and Marcelo Fernández-Lahore

1 Introduction

Downstream processing costs in biomolecule purification can constitute up to 80 % of the total manufacturing costs (Anspach et al. 1999). Increasing competition in the biotech markets and pressure from the regulatory agencies to keep the costs low triggered the development of novel and efficient (bio) separation technologies (DSouza et al. 2013). Process integration is one of the standard routines in the field of biochemical engineering. Process integration refers to the reduction of processing steps by combining multiple unit operations, which will in turn help reduce product and time losses incurred in individual steps. One of such integrated process approaches in downstream processing is Expanded Bed Adsorption (EBA), which has been already successfully applied in the purification of a variety of biomolecules (Anspach et al. 1999; de Sousa Junior et al. 2015). An illustration of the integrative approach is presented in Fig. 1.

In 1992 Chase and Draeger demonstrated a novel approach in which feedstock can be fed to the EBA system without earlier removal of particulate material by centrifugation or filtration, thus showing considerable potential for this approach in simplifying downstream processing steps (Draeger and Chase 1991). Performance of EBA mainly depends on the column hydrodynamics, which in turn depends on the stability of a perfectly fluidised bed in the presence of complex feedstock to obtain the best results. Cell adhesion on the adsorbent poses a threat to the success of EBA technology. It will result in formation of internal channels in the EBA bed,

V. Yelemane (✉) · M. Kangwa · M. Fernández-Lahore
Downstream Bioprocessing Laboratory, School of Engineering and Science Jacobs University, Campus Ring 1, 28759 Bremen, Germany
e-mail: v.yelemane@jacobs-university.de

Prasanna B.D. et al. (eds.), *Biotechnology and Biochemical Engineering*,
DOI 10.1007/978-981-10-1920-3_16

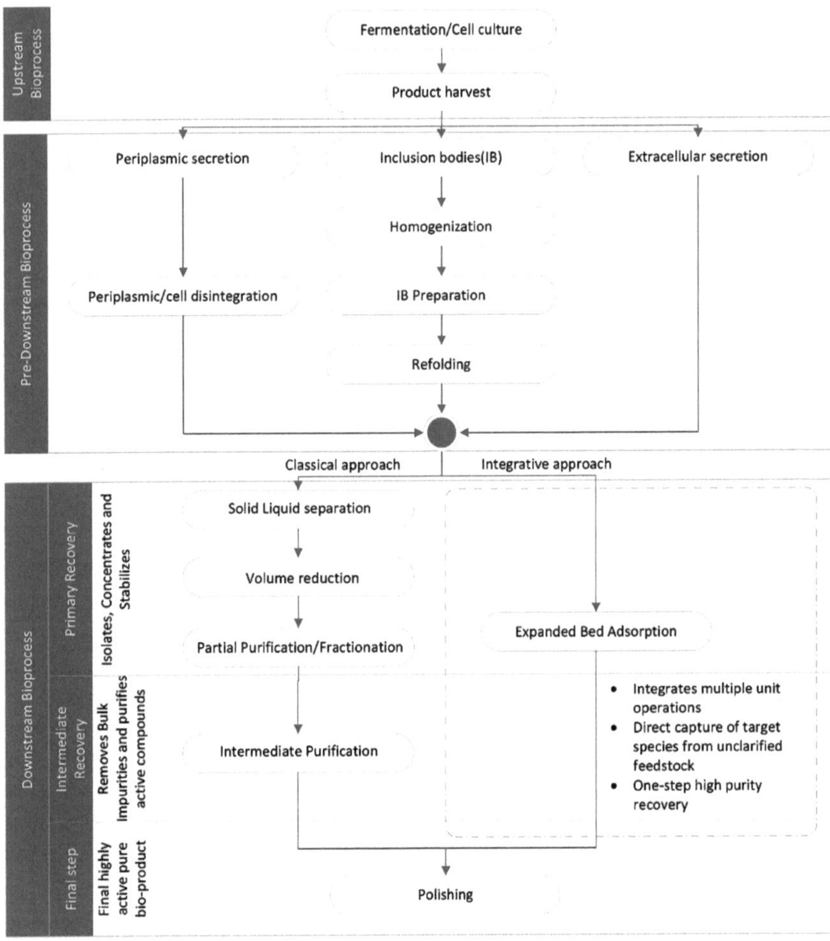

Fig. 1 Process flow for classical and integrative downstream processing

which leads to irregular colloidal distribution in the column resulting in bed collapse (Fernandez-Lahore et al. 2000). To alleviate such interaction, polymer shielding of the chromatographic adsorbents has been proven to be among the more successful approaches (Vennapusa and Fernandez-Lahore 2010). The main objective of this work is to develop simple microwell plate assay for high throughput screening polymer with cell repellent properties—mainly towards *Saccharomyces cerevisiae* (baker's yeast) cells. Validation experiments will be performed using microscopy techniques by coating the screened polymer onto EBA adsorbent to test its cell repellent property in process conditions.

2 Materials and Methods

2.1 Chemicals and Reagents

Polystyrene flat-bottom 96 microwell plate were purchased from Greiner Bio-One GmbH, Germany. Polymers polyvinyl alcohol (PVA), poly ethylene glycol (PEG 26), Triton X100, Brij 58, Tween 20, Polyvinylpyrrolidone (PVP360), DEAE-dextran, Glycidylmethacrylate (GMA), dimethylsulfoxide (DMSO) and 2,2′-Azobis(2-methylpropionitrile (AIBN) from Sigma–Aldrich, Germany. Poly acrylic acid from Polysciences, Germany. All other chemicals and salts were purchased from AppliChem, Germany. Milli-Q purified water was used.

2.2 Cell Culture

Saccharomyces cerevisiae cells were cultivated in yeast peptone dextrose medium in Erlenmeyer flasks on an orbital shaker set at 220 rpm and 30 °C. Cells were harvested at late log phase by centrifugation and washed with PBS, Milli-Q and 20 mM phosphate buffer at pH 7.5 (Eq. Buffer), in that order, respectively. For all experiments fresh cells, with more than 95 % viability, were used.

2.3 Epoxy Coating of Microwell Plate

Initially, Poly(glycidylmethacrylate) was synthesised by homo-polymerisation of glycidylmethacrylate in methanol with 2,2′-Azobis(2-methylpropionitrile) as initiator. Reaction mixture consists of 20 % v/v GMA and 1 % w/w AIBN in methanol. Reaction was initiated by heating the mixture to 70 °C for 30 min. The polymerisation reaction was stopped at gelling point by diluting with DMSO. PolyGMA was then dip coated on to 96 well plates. The residual solvent was evaporated at 40 °C for 16 h. Finally, the plates were washed thoroughly with Milli-Q water and dried at room temperature before use (Eckert et al. 2000).

2.4 Coupling of DEAE-Dextran to Epoxy Groups in Microwell Plate

10 mg/mL of DEAE-dextran was dissolved in reagent solution (1 mL of 1 M sodium hydroxide, 3.6 mg of sodium borohydride, and 0.4 mL of water). 100 μL of the solution was added per well for the coupling reaction. Plates were incubated for 18 h on plate mixer at room temperature. Then plates were washed with eq.

buffer till plate pH reaches neutral and also to remove unreacted residues. The plates are stored in same buffer at 4 °C.

2.5 *Polymer Coating and Cell Adhesion Studies*

300 μL per well of polymer solution with 1 % v/v in Eq. buffer was loaded on plate and incubated for 30 min at room temperature. The plate was then washed repeatedly with eq. buffer to remove the unbound polymer. Cell suspension of concentration one OD600 was taken in quantities of 300 μL per well and incubated for 3 h. After the incubation cell suspension was removed and 150 μL of 0.2 % w/v Crystal Violet dye in water was added and plates were incubated for 15 min. Plate was gently washed 2–3 times with 300 μL of Eq. buffer to remove excess dye and unbound cells. Subsequently, the plate was dried overnight. The dye was eluted with 150 μL of 30 % acetic acid by incubating plate on a plate mixer for 15 min. The eluted mixture was transferred to a fresh microwell plate and absorbance was measured at 570 nm.

3 Results and Discussion

In a previous study, we have developed agarose gel based discs, which can be functionalised with standard ligands in easy steps and placed in microwell plates for similar studies. However, the main limitation of this method was the difficulty in the estimation of cell adhesion. It was found that the gel adsorbed crystal violet dye which resulted in a false positive result when testing. To overcome this challenge current method has been developed.

3.1 *PolyGMA Coating and Dextran Coupling*

In the first stage, PolyGMA was synthesised as mentioned previously. During the polymerisation reaction the acrylic part of GMA plays the main role and polymerises—resulting in freely accessible epoxide groups in the polymer. Coating with epoxy groups is an effective and easy way to modify inert surfaces. The layer of polyGMA will provide free epoxy functional groups which can then be easily utilised for further coupling of proteins or polymers using well proven techniques. Last well of the every row was left without coating and later used as negative control. In second stage, polyGMA coated plates were coupled with commercially available functionalised dextran. In this study we used DEAE-dextran, a functionalised polysaccharide similar to DEAE-agarose used in expanded bed adsorption. Protocol mentioned in methods was used for coupling. Wide variety of functionalised dextran with varying ionic and hydrophobic properties, which mimic chromatographic conditions, are commercially available. Using such a product saves a significant amount of time. At end of the coupling reaction plates were washed 2–3 times to remove any uncoupled material and stored in eq. buffer. Figure 2 represent modification process graphically (Fig. 3).

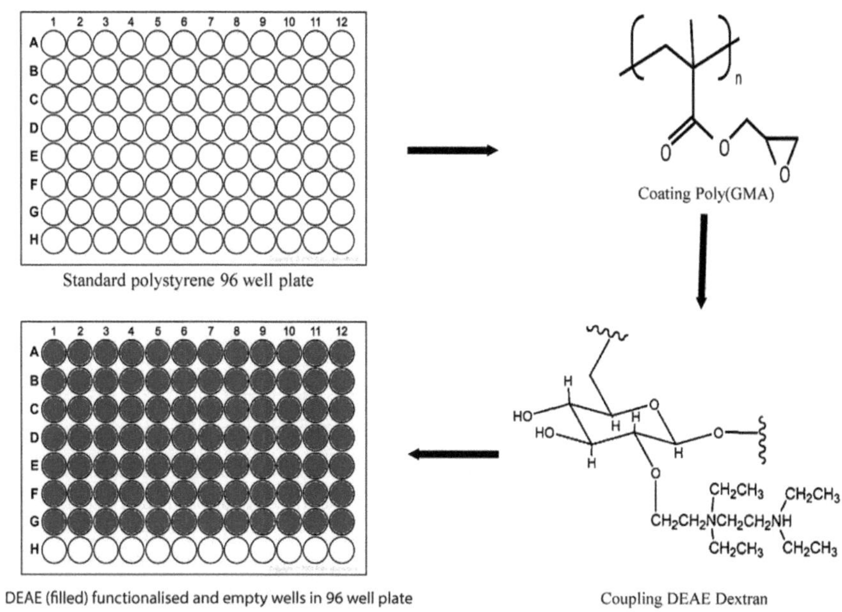

Fig. 2 Schematic representation of step by step modification of microwell plate

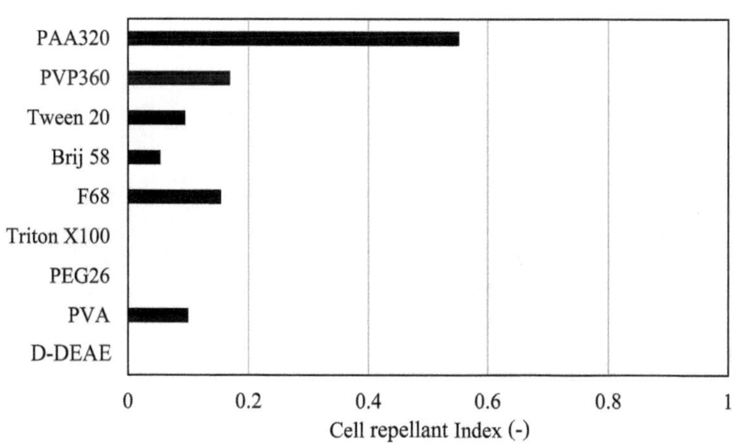

Fig. 3 List of polymer tested and their cell repellant index for yeast cells

3.2 Effect of Polymer Coating on Cell Adhesion

In this study, we have chosen a set of ionic and non-ionic polymers with high molecular weight viz., polyvinyl alcohol, poly ethylene glycol, Triton X100, Brij 58, Tween 20, Polyvinylpyrrolidone and polyacrylic acid for coating. Polymer coating was performed as described in the previous section. Native or non-coated and empty wells were used as positive and negative control respectively. Polymer coating modified the surface properties of DEAE-dextran and this change influence cell-surface interaction. To measure such change in interaction, crystal violet (CV) assay was used and performed as described in the previous section. CV binds to cells and is a well proven method to study biofilms because the dye binds proportionally to cell number. Post staining, plates were washed 2–3 times to remove any unbound cells or excess stain and dried overnight. The colour intensity at bottom of the plate will give qualitative estimation of cell adhesion. We employed dye elution techniques where, CV was eluted with acetic acid to estimate the cell adhesion quantitatively. Results of the cell adhesion assay were expressed as dimensionless cell repellent index, which is defined as

$$\text{Cell repellent index} = 1 - \frac{0D_{\text{coated}}}{0D_{\text{Neat}}} \quad (1)$$

Results showed highest interaction with native Dextran-DEAE surface. PEG and triton X100 coating did not have any influence on the cell adhesion. Pluronic F68 and PVP 360 increased the cell repulsion property by 15 % but highest cell repulsion property was demonstrated by PAA 320 with more than 50 %. To validate these results, PAA 320 was coated on Streamline DEAE, a commercial EBA adsorbent and tested with partition experiments followed by microscopic observation for cell adhesion. Figure 4 shows the microscopy images of the adsorbent

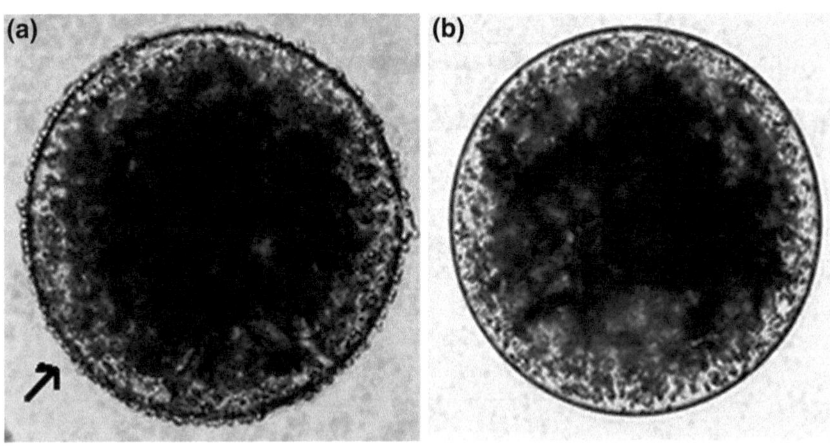

Fig. 4 Deposition of yeast cells *S. cerevisiae* onto Streamline DEAE. **a** Interaction of cells (*arrow*) on to the naked adsorbent. **b** Reduced yeast cell interaction onto PAA coated adsorbent

after the adhesion experiments figure (A) shows the naked bead with a clear interaction with cell and forming a thin layer on the bead surface. Figure (B) Shows PAA coated bead showing cell repulsion property showing very negligible interaction exists at the adsorbent surface. The interaction between the substrate and yeast needs to be studied further. Microscopic visualisation clearly demonstrate the cell repulsion property of PAA 320 and it corroborates the results of microwell plate method. It demonstrating that polymer coating can reduce cell adhesion in integrated bioprocessing and enhance process efficiency.

4 Conclusion

In conclusion, this paper demonstrates a robust and simple method for microwell plate assay for cell adhesion studies. It can be used for high-throughput screening cell repulsion polymers. Plate surface can be modified with protein or polymers by coating of grafting techniques to fit our requirement. In this study we tested eight polymers for cell repulsion property. PAA, a pharmaceutical grade polymer interacted with plate surface and adsorbent. We assume that it formed a layer on substrate and altered the surface properties. It was directly indicated by change is cell adhesion behaviour after coating. The optimal combination found from the experiments performed was used to coat the commercial adsorbent for further tests. Coated adsorbents showed similar cell repulsion property in process like conditions and were validated using microscopic visualisation.

Acknowledgments The authors gratefully acknowledge financial support from the European Union Seventh Framework Programme (EU-FP7 KBBE 2012 Project 312004-INTENSO).

References

Anspach, F.B., Curbelo, D., Hartmann, R., Garke, G., Deckwer, W.-D.: Expanded-bed chromatography in primary protein purification. J. Chromatogr. A **865**, 129–144 (1999)

de Sousa Junior, F.C., Vaz, M.R.F., de Araújo Padilha, C.E., Chibério, A.S., Martins, D.R.A., de Macedo, G.R., dos Santos, E.S.: Recovery and purification of recombinant 503 antigen of *Leishmania infantum* chagasi using expanded bed adsorption chromatography. J Chromatogr. B **986–987**, 1–7 (2015)

Draeger, N.M., Chase, H.A.: Liquid fluidized bed adsorption of protein in the presence of cells. Bioseparation **2**, 67–80 (1991)

DSouza, R.N., Azevedo, A.M., Aires-Barros, M.R., Krajnc, N.L., Kramberger, P., Carbajal, M.L., Grasselli, M., Meyer, R., Fernández-Lahore, M.: Emerging technologies for the integration and intensification of downstream bioprocesses. Pharm. Bioprocess. **1**, 423–440 (2013)

Eckert, A.W., Grobe, D., Rothe, U.: Surface-modification of polystyrene-microtitre plates via grafting of glycidylmethacrylate and coating of poly-glycidylmethacrylate. Biomaterials **21**, 441–447 (2000)

Fernandez-Lahore, H.M., Geilenkirchen, S., Boldt, K., Nagel, A., Kula, M.R., Thommes, J.: The influence of cell adsorbent interactions on protein adsorption in expanded beds. J. Chromatogr. A **873**, 195–208 (2000)

Vennapusa, R.R., Fernandez-Lahore, M.: Effect of chemical additives on biomass deposition onto beaded adsorbents. J. Biosci. Bioeng. **110**, 564–571 (2010)

Concentration of *C-Phycocyanin* from *Spirulina platensis* Using Forward Osmosis Membrane Process

Shoaib A. Sharief and Chetan A. Nayak

1 Introduction

Phycobiliproteins are a gathering of water soluble colored proteins that are ordinarily found in red algae and Cyanobacteria. They have a wide range of uses and are known for their fluorescent application in the field of immunological and clinical analysis. Their additional uses incorporate as a food colorant and also as a therapeutic agent (Sekar and Chandramohan 2008). Phycobilisome are comprised of *allophycocyanin* centers that are fenced by *phycocyanin* at the fringe. The pivotal part is *phycocyanin* though the connecting pigment between photosynthetic lamella and phycobilisomes is the *allophycocyanin* (Eisele et al. 2000). The cyanobacteria to be specific, *Spirulina platensis* have been promoted in couple of countries in view of its critical constituents, for example, vitamins and proteins. *Spirulina platensis* is a cheap and rich wellspring of color like *phycocyanin* (Kato 1994; McCarty 2007). It is likewise utilized as a part of food colorant in products, for example, yogurt, chewing gum, mixed beverages soda pops, milk shakes beauty care, furthermore in pharmaceutical (Babu et al. 2006). Water is the significant constituent in a large portion of natural color extract that helps in development of microorganism. Shelf life and usage of these extracts can be increased by elimination of water by concentrating and freeze drying. *Phycocyanin* is a heat sensitive protein (Sarada et al. 1999; Chethana et al. 2015). The industrial processing of the food substances has a tremendous and serious impact on the final product as it affects the nutritional as well as sensorial properties of the product. In order to reduce this impact, membrane technology can be used as an optional procedure to concentrate the *phycocyanin*. For the concentrating of natural colorant, non-heat sensitive procedures like Forward Osmosis (FO) is been reported (Nayak et al.

S.A. Sharief · C.A. Nayak (✉)
Department of Chemical Engineering, BMS College of Engineering,
P B NO. 1908 Bull Temple Road, Bengaluru 560019, Karnataka, India
e-mail: canayak.che@bmsce.ac.in

Prasanna B.D. et al. (eds.), *Biotechnology and Biochemical Engineering*,
DOI 10.1007/978-981-10-1920-3_17

2010a, b; Rastogi and Nayak 2010). The grade or the quality of food is maintained by FO, without bringing about any changes to the physical properties, (for example, smell, shading, nourishment and taste) (Nayak et al. 2011; Sant' Anna et al. 2012). It uses a hydrophilic semi-permeable membrane, which separates the osmotic and feed solution having particular osmotic pressure. The main driving force is osmotic pressure difference across the membrane. The objective of this present work is to extensively study the Forward osmosis process for concentrate the purified phycocyanin biomolecule.

2 Materials and Methods

2.1 *Organism and Chemicals*

The blue green algae (*Spirulina platensis*) in the dried powder form was kindly provided by M/s. Parry Nutraceuticals, a Division of E.I.D Parry (India) Ltd., Chennai. Sodium Chloride (NaCl) and Sodium phosphate was procured from S.D Fine chemicals, India. Sodium Dodecyl Sulfate (SDS) was procured from Sisco research lab, India. All the chemicals used were of analytical grade.

2.2 *Extraction and Purification of C-PC from* Spirulina platensis

The Phycobiliprotein extraction involves the rupture of cell biomass for the extraction of protein from the cell of *Spirulina platensis*. The pigment C-PC was extracted from *Spirulina platensis* by using techniques like: (a) Water extraction: Dry *Spirulina platensis* biomass was suspended in distilled water and maintained at room temperature for 2 h and the C-PC leached out was centrifuged at 10,000 rpm at 4 °C for 10 min. (b) Homogenization of cells using Homogenizer: Wet Biomass was homogenized by Growell Homogenizer from (Bio-Lab, Growell instruments Pvt. Ltd. India) at 5000 rpm for 10 min and the extract was centrifuged at 10,000 rpm at 4 °C, supernatant was collected. (c) Freezing and thawing: C-PC was also extracted by repeated freezing and thawing of wet biomass cells in 50 mM phosphate buffer pH 6.8 for comparison. Deep freezing at temperature of −69 °C was carried out for 2 h, immediate thawing was performed. The extract was centrifuged at 10,000 rpm. At 4 °C for 10 min and the supernatant contained C-PC.

To separate and eliminate small cell fragments in the crude C-PC extract, Whatman filter paper of Grade 1:11 μm (medium flow filter paper) was used in the coarse filtration step. The filter paper was placed on a funnel and the crude C-PC sample taken after extracting using homogenization method was used. For the fine filtration step, Tangential Flow Filtration (TFF) unit (Millipore, Pellicon XL with

PS pump, USA) was modified for Microfiltration. Module I was connected to the TFF unit. Polypropylene membranes (Sterlitech corporation, USA) disc of 0.047 m diameter with a pore size 0.45 μm, was placed between the module I plates and concealed by connecting it to the peristaltic pump (Master flex L/S Cole Parmer, USA.) The feed was loaded into the feed tank to check the transmembrane flux. The feed transmembrane pressure was maintained at 1.4 kg/cm^2 and the retentate pressure was maintained at 1.3 kg/cm^2 this experiment conducted for a period of one hour. Periodically, the membrane was changed in order to increase the transmembrane flux and to avoid fouling. The final pure filtrate was obtained by using syringe filtration in batch process. The complete feed is forced through the Polypropylene membrane (Sterlitech Corporation, USA) disc of 0.025 m diameter and pore size 0.2 μm to get the final product.

2.3 Feed and Osmotic Agent

Crude C-PC and purified C-PC was used as feed by using different modules each time. Double distilled water was used as feed as into compare the transmembrane flux with that of the C-PC. OA solution were prepared by dissolving sodium chloride in distilled water in various degrees of (1.0–6.0 M) concentrations. These solutions were kept overnight at room temperature before utilization to ensure complete dissolution of sodium chloride.

2.4 Experimental Setup and Membrane Modules for Concentration of C-PC

Two acrylic membrane Module I (Fig. 1a. Circular module) and Module II (Fig. 1b. Flat membrane module) were fabricated and used for study which is represented in Fig. 1. For the concentration of feed by FO process, the experimental setup consisted of a hydrophilic, semi permeable Cellulose Tri Acetate (CTA) membrane (HTI Technologies, USA), placed between the modules. Feed solution (water, C-PC extract) and osmotic agent (1.0–6.0 M NaCl) was passed on either sides of the membrane in co-current direction by using peristaltic pumps. The effect of change in feed and OA flow rate (25–125 mL min^{-1}) on transmembrane flux was studied. Transmembrane flux was calculated by measuring the increase in volume of OA for every 1 h. Small scale experiment was performed for 3 h by using module I and the average values of the flux was reported. Large scale experiment was carried out for a period of 11 h by using module II. Reverse solute diffusion was estimated by using flame photometer (Systronic flame photometer Model No. 128). All the experiments were performed at room temperature of 25 ± 2 °C.

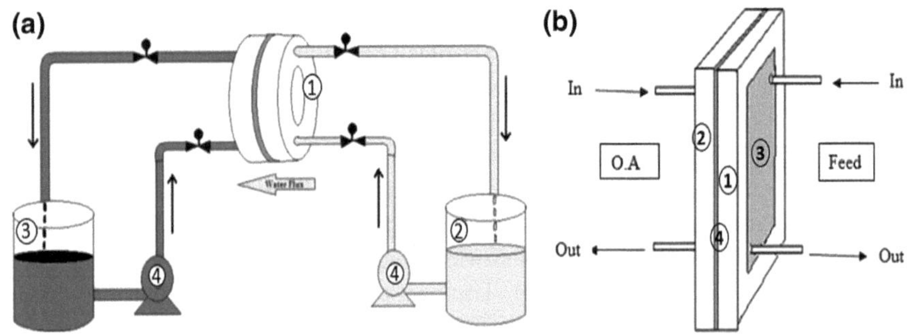

Fig. 1 Illustrative diagram of (**1a**) *1* Circular membrane module I for FO process: *1* circular membrane module of membrane area of 0.00519 m^2, *2* feed reservoir, *3* OA reservoir and *4* peristaltic pump. **1b** Flat membrane module II-Front acrylic module. *2* Back acrylic module. *3* Membrane area of 0.012954 m^2. *4* Silicon rubber gasket

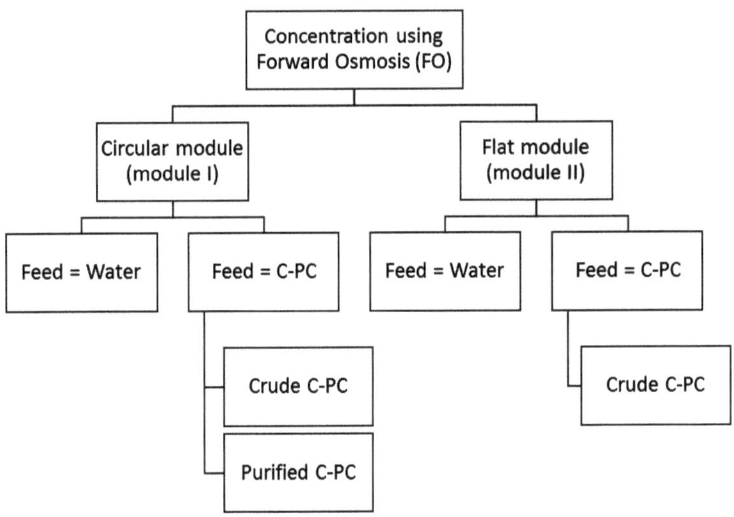

Fig. 2 Flow chart of forward osmosis membrane process

The process of concentration using FO membrane process is depicted in Flow chart Fig. 2.

2.5 *Determination of C-Phycocyanin Concentration*

The concentration of C-Phycocyanin was determined by using a UV-Visible spectrophotometer (Model: Cary 8454, Agilent Technologies, Inc.) by measuring the absorbance at 620 nm. Digital refractometer (HI 9680, HANNA) was used in estimation of Total soluble solid (TSS) in Brix. The C-Phycocyanin concentration was calculated using the following Eq. 1 as given by (Siegelman et al. 1978; Chethana et al. 2015).

$$\mathrm{C-PC}\ (\mathrm{mg\ mL^{-1}}) = \frac{[A_{620} - 0.474(A_{652})]}{5.34}. \quad (1)$$

where A_{620} is the Absorbance of C-PC and A_{652} is the absorbance of Allophycocyanin.

3 Results and Discussion

3.1 Effect of Change in Osmotic Agent Concentration on Transmembrane Flux

Experiments were conducted using module I. On account of water and C-PC as a feed, the transmembrane flux was found to increase from 1.978 to 2.939 $\mathrm{lm^{-2}\ h^{-1}}$ and 0.614 to 1.095 $\mathrm{lm^{-2}\ h^{-1}}$ with an increase concentration of OA from 1.0 to 6.0 M NaCl as shown in Fig. 3.

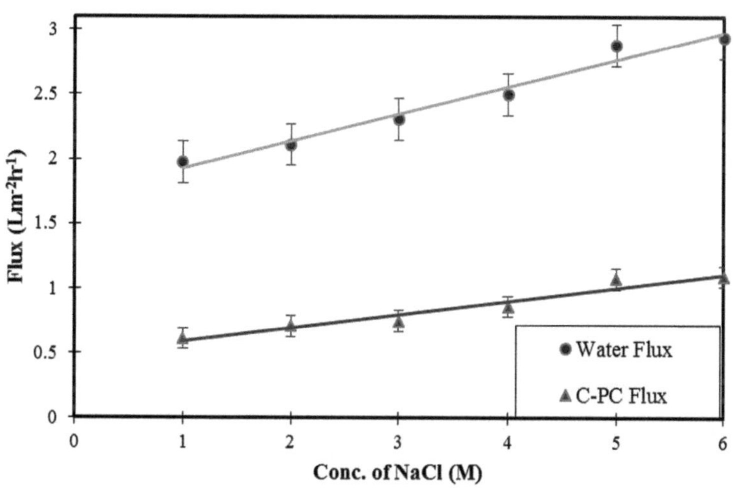

Fig. 3 Variation of osmotic agent concentration versus feed transmembrane flux

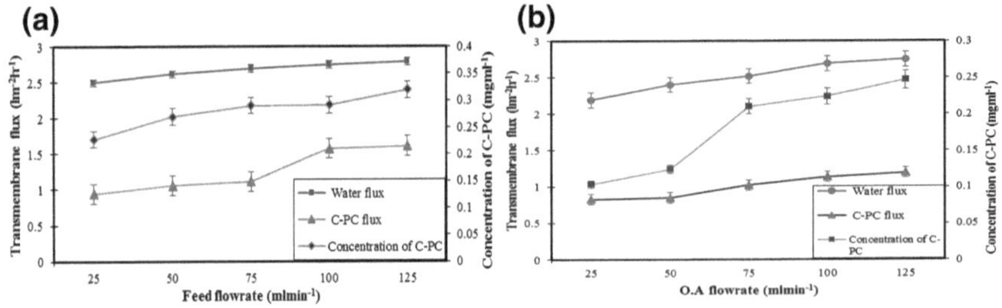

Fig. 4 **a** Effect of change in feed flow rate on transmembrane flux and on concentration of C-PC. **b** Effect of change in osmotic agent flow rate on transmembrane flux and on concentration of C-PC

3.2 Effect of Change in Feed and Osmotic Agent Flow Rate

With an increase in feed flow rate from 25 to 125 mL min^{-1}, the transmembrane flux for water and C PC was found to increase from 2.498 to 2.786 lm^{-2} h^{-1} and 0.941 to 1.614 lm^{-2} h^{-1} respectively Phycocyanin concentration was found to increase from 0.228 to 0.320 mg mL^{-1} Fig. 4a. The TSS in the C-Phycocyanin sample increased from 3.1 °Bx to 4.9 °Bx respectively. The transmembrane flux was found to increase from 0.826 to 1.191 lm^{-2} h^{-1} the C-Phycocyanin concentration was calculated by using the Eq. 1 and it was found to increase from 0.1031 to 0.247 mg mL^{-1} Fig. 4b. The TSS in the C-Phycocyanin sample increased from 3 °Bx to 5.1 °Bx, respectively.

3.3 Large Scale Experiments Using Module II

Based upon the optimized conditions, the crude C-PC content was concentrated from 0.288 mg mL^{-1} initial concentration to 0.313 mg mL^{-1} final concentration 8.68-fold increase in concentration. Due to increase in feed concentration, the transmembrane flux was found to decrease from 0.602 to 0.502 lm^{-2} h^{-1} with time. Babu et al. (2006) have reported a maximum concentration of 3-fold for Phycocyanin for osmotic membrane distillation process, wherein 2.9 °Bx crude extract was concentrated up to 5.3 °Bx C-PC concentrate.

3.4 Purification of C-PC and Concentration for 11 h

The purification of CPC was carried out; the total volume and the concentration of C-PC found in the crude extract were considered to be 100 %, during coarse filtration when the sample was filtered through a membrane of pore size 1 μm, the reduction in volume from 100 to 85.1 %. Later, after the process of fine filtration i.e., by using microfiltration and syringe filtration wherein membranes of pore size

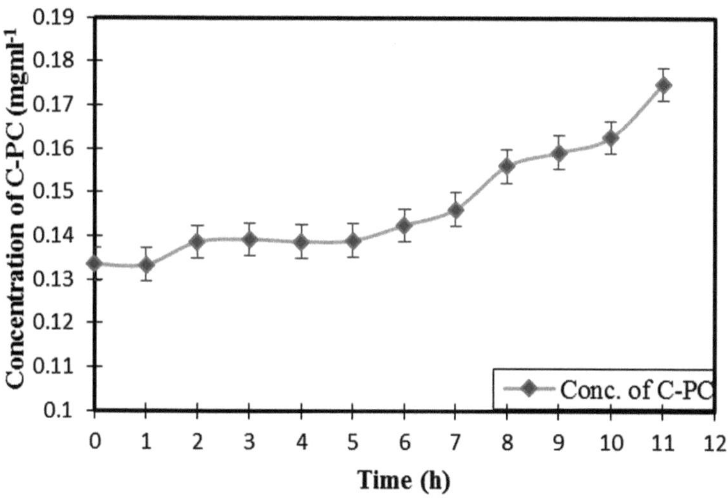

Fig. 5 Variation of concentration of purified C-PC during large scale experiment using module I

0.45 and 0.2 μm respectively were used. The volume of the C-PC extract was further reduced to 76 %. There was a change in the color of the extract changed from dark blue to blue after fine filtration processes However, the purity ratio of the supernatant liquid was slightly increased. When C-PC extract was filtered through membrane with 0.45 μm pore size, an average level of permeate flux of 1.304 lm^{-2} h^{-1} was determined. After nearly 70 min of the microfiltration process there was reduction seen in the permeate flux, this may be attribute to membrane fouling. Later, the membrane was replaced in order to improve the permeate flux and to complete the filtration process. After the microfiltration process the purified C-PC permeate was subjected to dead end filtration by the help of syringe filters. Sterile Cellulose Acetate membranes of pore size 0.2 μm wherein the ultimate fine filtration was carried out. The result of syringe filtration was a purified C-PC and the purity ratio (A_{620}/A_{280}) was also determined. There was an increase from 0.2977 of crude C-PC to 0.597 purified C-PC. During FO, Initial volume of 99.3 mL of C-PC was reduced to 70.6 mL. Simultaneously, C-Phycocyanin content was concentrated from 0.133 mg mL^{-1} initial concentration to 0.175 mg mL^{-1} final concentration (Fig. 5) it was also found that around 2.9 °Bx initial C-PC extract was concentrated to 5.3 °Bx C-Phycocyanin concentrate. To the purity ratio increasing from initial crude C-PC of 0.2977 to final pure C-PC of 0.597 (Fig. 6).

3.5 Confirmation of C-PC Using SDS-PAGE

The purity of C-PC sample obtained after syringe filtration and FO were confirmed by using SDS-PAGE. SDS-PAGE indicates the majority of C-PC (α and β) subunits

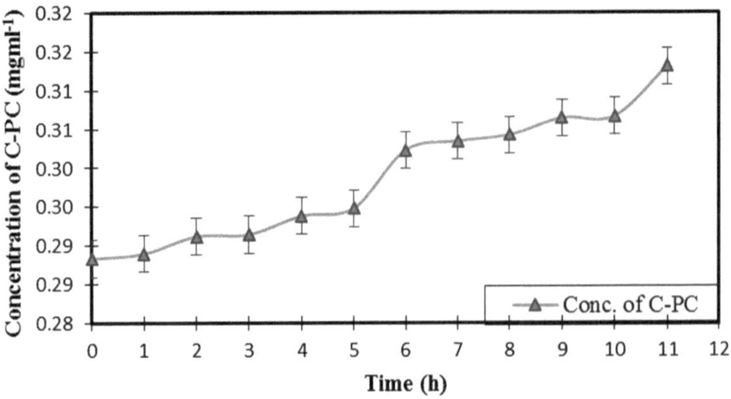

Fig. 6 Variation of C-PC conc. in large scale experiment using module II

Fig. 7 SDS-PAGE of C-Phycocyanin. *Lane-1* molecular marker; *lane-2* crude C-PC + FO; *lane-3* pure C-PC + FO

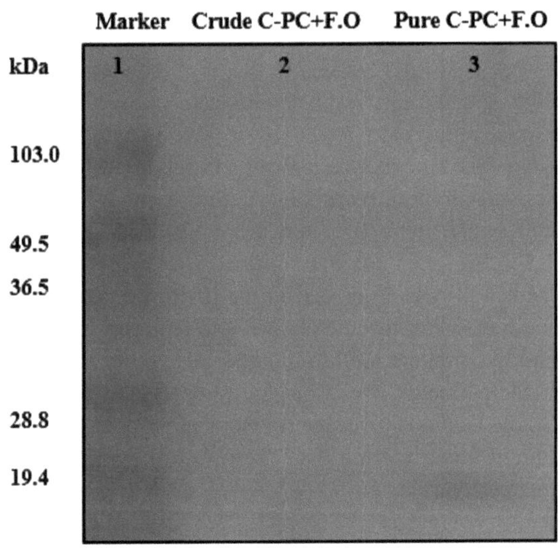

concentration at around 19.4 kDa in Lane-3 which is obtained after Syringe filtration and FO process as shown in Fig. 7.

4 Conclusion

Forward osmosis membrane process for concentrate the purified phycocyanin was studied. The effect of change in feed and OA flow rate was found prominent at higher concentration of the phycocyanin. The crude C-PC content was concentrated from an initial concentration of 0.288 to 0.313 mg mL^{-1} using module II. The pure C-PC obtained by microfiltration using syringe showed a change in concentration from an initial concentration of 0.133 to 0.175 mg mL^{-1} by using module I for FO. The purity ratio was also found to increase from initial crude C-PC of 0.2977 to final purified C-PC of 0.598 which is nearly equal to a food grade C-PC. The forward osmosis membrane process is a promising method for concentration of value added biomolecules.

Acknowledgments The work reported in this paper is supported by the college through the TECHNICAL EDUCATION QUALITY IMPROVEMENT PROGRAMME [TEQIP-II] of the MHRD, Government of India. Authors express their sincere thanks to Vision Group of Science and Technology (VGST). This project is continuation of the project granted by (VGST), Karnataka State to Dr. Chetan A Nayak, Project grant No. GRD 204 for 2013–14.

References

Babu, B.R., Rastogi, N.K., Raghavarao, K.S.M.S.: Mass transfer in osmotic membrane distillation of phycocyanin colorant and sweet-lime juice. J. Membr. Sci. **272**, 58–69 (2006)

Chethana, S., Nayak, C.A., Madhusudhan, M.C., Raghavarao, K.S.M.S.: Single step aqueous two-phase extraction for downstream processing of C-phycocyanin from *Spirulina platensis*. J. Food Sci. Technol. **52**(4), 2415–2421 (2015)

Eisele, L.E., Bakhru, S.H., Liu, X., MacColl, R., Edwards, M.R.: Studies on C-phycocyanin from *Cyanidium caldarium*, a eukaryote at the extremes of habitat. Biochim. Biophys. Acta **1456**(2–3), 99–107 (2000)

Kato, T.: Blue pigment from *Spirulina*. New Food Ind. **29**, 17–21 (1994)

McCarty, M.F.: Clinical potential of Spirulina as a source of phycocyanobilin. J. Med. Food **10**(4), 566–570 (2007)

Nayak, C.A., Rastogi, N.K.: Forward osmosis for the concentration of anthocyanin from *Garcinia indica* Choisy. Sep. Purif. Technol. **71**, 144–151 (2010a)

Nayak, C.A., Rastogi, N.K.: Comparison of osmotic membrane distillation and forward osmosis membrane processes for concentration of anthocyanin. Desalin. Water Treat. **16**, 134–145 (2010b)

Nayak, C.A., Valluri, S.S., Rastogi, N.K.: Effect of high or low molecular weight of components of feed on transmembrane flux during forward osmosis. J. Food Eng. **106**, 48–52 (2011)

Rastogi, N.K., Nayak, C.A.: Membranes for Membranes for forward osmosis in industrial applications. In: Basile A., Nunes S.P. (eds.) Advanced Membrane Science and Technology for Sustainable Energy and Environmental Applications, pp. 680–717. Woodhead Publishers, Cambridge (2010)

Sarada, R., Pillai, M.G., Ravishankar, G.A.: Phycocyanin from *Spirulina* sp: influence of processing of biomass on phycocyanin yield, analysis of efficacy of extraction methods and stability studies on phycocyanin. Process Biochem. **34**, 795–801 (1999)

Sekar, S., Chandramohan, M.: Phycobiliproteins as a commodity: trends in applied research, patents and commercialization. J. Appl. Phycol. **20**(2), 113–136 (2008)

Siegelman, H.W., Kycia, J.H.: Algal biliproteins. In: Hellebust J.A., Craigie J.S. (eds.) Handbook of Phycological Methods, pp. 72–80. Cambridge University Press, Cambridge (1978)

Sant' Anna, V., Marczak, L.D.F., Tessaro, I.C.: Membrane concentration of liquid foods by forward osmosis: process and quality view, J. Food Eng. **111**, 483–489 (2012)

Design and Fabrication of Miniature Bubble Column Bioreactor for Plant Cell Culture

K. Sandesh, P. Ujwal, Blecita D. Mascarenhas, Gayatri Dhamannavar, Narmada Kumar and Dakshayini

1 Introduction

The bubble column bioreactor is an apparatus having its shape in the form of a column. The aeration is introduced from the base of the column which keeps the reaction medium in its mixed form (Deckwer and Schumpe 1993). The various features like low capital costs, simple mechanical configurations, low energy consumption and a minimum operational cost makes it an efficient system to be implemented for any product operation (Betts et al. 2006).

Takayama and Misawa (1981) reported for the first time about the use of bioreactor for micropropagation in Begonia. As in our traditional method, the task of plant cell culture techniques (Bányai et al. 2003) requires many sensitive protocols and the maintenance of the entire propagation needs routine monitoring. Considering the same, its specific that micropropagation by this technique is labour intensive. Bioreactors based propagation can increase the multiplication rate, culture growth and space reduction (Karppinen et al. 2006). It minimizes the expenditure of energy and manual operational protocols in other commercial purposes. These bioreactors can be a key step to produce secondary metabolites (Choi et al. 1990) in large scale through plant cell cultures, organ cultures and hairy root cultures (Saurabh et al. 2002). Controlled production conditions, product yield variations and high quality product can be maximized in these bioreactors thus simplifying the various processes (Paek et al. 2005 and Ozlem et al. 2010).

In the present study, the bioreactor constructed provided optimum condition, with respect to temperature, pH and basic function. The functional bioreactor after

K. Sandesh (✉) · P. Ujwal · B.D. Mascarenhas · G. Dhamannavar · N. Kumar · Dakshayini
Department of Biotechnology Engineering, NMAM Institute of Technology
(Affiliated to VTU Belgaum), Nitte, Karkala Taluk, Udupi, Karnataka, India
e-mail: sandeshk@nitte.edu.in

Prasanna B.D. et al. (eds.), *Biotechnology and Biochemical Engineering*,
DOI 10.1007/978-981-10-1920-3_18

construction was studied with the biomass of plant cell callus of *Cichorium intybus* (Chicory). A batch wise operational mechanism was observed for a required period of time and the yield of the cell mass was calculated.

2 Materials and Methods

2.1 Preparation of Callus

The explants from the *C. intybus* species was collected, surface sterilised and cultivated in Murashige and Skoog (MS) semisolid medium supplemented with 1 g/L benzylaminopurine (BAP) and 1 g/L Naphthalene acetic acid (NAA) (Nandagopal et al. 2007). The callus was grown in a span of 15–20 days.

The callus from the semi solid media was removed and finely chopped using sterile knife in the laminar air flow. The Petri plate containing the callus was sealed with paraffin tape and weighed. 2 g of this callus was inoculated into the shake flask and bubble column bioreactor.

2.2 Shake Flask Study

The growth of *C. intybus* callus was carried out in 2 L conical flask kept in rotary shaker under the source of light. Since the growth of callus differs from one flask to another, scaling up becomes difficult. Thus to overcome this problem, inoculation of the callus weighing 2 g was added to a 2 L of conical flask and kept in rotary shaker. Contamination was observed after 2 days of inoculation.

2.3 Designing of Bubble Column Reactor

The reactor wall of the bubble column bioreactor was made of glass material with the aspect ratio 2, the column length 22 cm and diameter 11 cm (Pirdashti and Kompany 2009). The callus was held under suspension by continuously sparing the sterile air through sparger consisting of evenly spaced 6 number of hole with 0.3 cm diameter at the bottom of the reactor (

Klass et al. 1991). The sterility of air was maintained by passing air through 0.45 μm filter (Fig. 1).

The content inside the bioreactor was autoclaved using electric heater. The sample port was provided with plunger to draw the sample at different time intervals. An outlet of 2 cm diameter was provided at the bottom of the reactor to drain out the contents at the end of the batch. Necessary monitoring devices and ports such as temperature, pH, light source, feed inlet and vent were provided on the head plate. Light source required to maintain photoperiod of 16 h for the growth of

Fig. 1 Bubble column bioreactor setup

callus was provided by LED lights equipped with time controller mounted inside the reactor head plate. It was set in such a way that the light was in ON mode for 16 h and in OFF mode for the next 8 h.

2.4 Parameter Estimation for Bubble Column Reactor

The calculation of Superficial gas velocity, circulation time and velocity of liquid at the center of the column and hold up was done by using the Eqs. (1)–(4) (Klass et al. 1991 and Bouaifi et al. 2001). The values obtained for the designed reactor setup is listed in Table 1.

The tower diameter, Tv = 11 cm and Height Hv = 22 cm

Superficial Gas Velocity,

$$V_{gc}^{c} = \frac{Q}{A} \tag{1}$$

Liquid Velocity at the Centre of the Column,

$$V_{lc} = 0.9(g * T_v * V_{gc}^{c})^{0.33} \tag{2}$$

Table 1 Parameter estimation for bubble column reactor

	Volumetric flow rate Q, L/min	Superficial gas velocity V^c_{gs} m/s	Liquid velocity V_{lc} m/s	Gas hold up ε	Circulation time (s)
Batch 1	3	1.18	0.97	0.67	2.28
Batch 2	2	0.79	0.85	0.51	2.61
Batch 3	1	0.39	0.68	0.31	3.28

Circulation Time, when $H_v > T_v$,

$$t_c = 11 * \frac{H_V}{T_V} \left(g * V^c_{gc} T_V^{-2}\right)^{-0.33} \tag{3}$$

Gas Hold Up for heterogeneous, When $d_b = 3$ mm
Hold up,

$$\varepsilon = 0.6\left(V^c_{gc}\right)^{0.7} \tag{4}$$

2.5 *Bubble Column Study*

The 2L volume of media was fed into the bubble column bioreactor. The media was autoclaved at 120 °C (Nandagopal et al. 2007). Once the media was cooled to room temperature 2 g of callus was inoculated into bioreactor by opening the butterfly valve at the top of the head plate. The sterile oxygen required for the growth of callus was supplied through the sparger and air flow rates were regulated by control valve. The experiment was carried out in 3 batches by varying the volumetric flow rate (Q) of air.

3 Results and Discussion

3.1 *Shake Flask Study*

In shake flask study the callus was inoculated to grow for 15 days. At the end of 15 days, *C. intybus* callus grown conical flask showed the maximum increase in biomass of 2 g in MS suspension media, with partial contamination 30–40 % of the flasks in every batch. The yield obtained in shake flask is shown in Fig. 2. Thus the mass of culture enhancement was drastically slow and no further growths of cells were observed in these flasks.

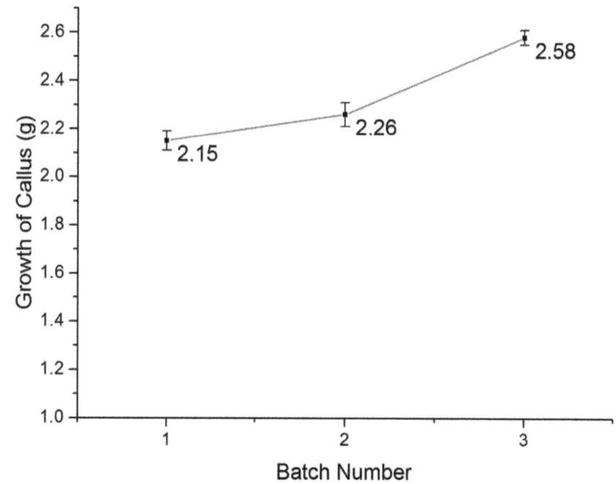

Fig. 2 Mass of callus grown in shake flask study

3.2 *Reactor Study*

A Preliminary study was carried out using callus cells in the bioreactor for determining the total biomass. As there are no much existing reports available with the study of bubble column bioreactors with the use of cell culture biomass. In the present investigation the experiment was carried out in 3 batches. In each batch the growth was efficient in bubble column bioreactor, since it provided all necessary conditions for the growth of plant cell and maintenance of sterility. As shown in Fig. 3 at the end of 15 days, the callus grown in 2 L bioreactor showed a maximum increase of 6.26 g in MS suspension media.

The yield obtained in batch 1, was comparatively less when compared to batch 2 and batch 3 in the reactor. The reason behind this could be the high air flow rate which resulted in foaming. Initially the air flow rate was maintained at 3 L/min which resulted in vigorous agitation and thus rupturing of callus. The initial flow

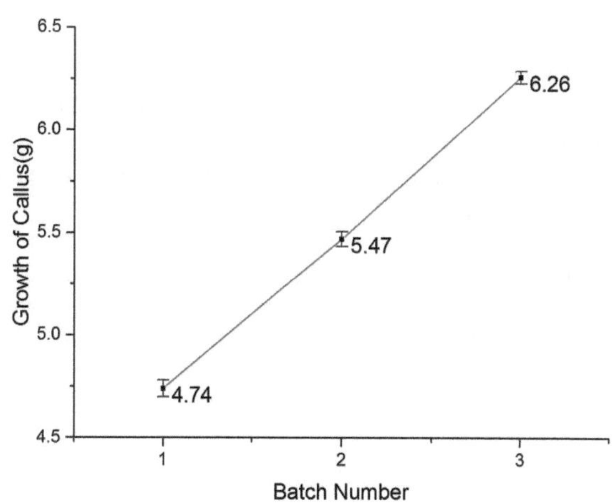

Fig. 3 Mass of callus in bubble column bioreactor

rate of 3 L/min resulted in foaming. This was observed after 8 days of inoculation. Due to agitation or fermentation an accumulation of fine frothy bubbles were formed surrounding the surface of the liquid. In the work done by Ozturk et al. (1987) on *Atropa belladonna,* the development of foam and maintenance of stability was investigated as a main function of culture conditions (Hyndman et al. 1997). In determining the total volume of foam, the rate of air flow was found to be of major importance and thus the highest foam levels were measured 11 days after inoculation (Behkish et al. 2002).

4 Conclusion

In the present study the construction of bubble column bioreactor for the production of commercially valuable secondary metabolite in vitro was discussed. Also, the observation of sterile environment without any contamination till 15 days with growth yield inside the reactor was a remarkable observation. The growth of callus was found to be 6.26 g/L, when compared to shake flask 2.58 g/L. This could prove that these bioreactors have a tremendous potential for the commercial scale synthesis of plant cell culture and their necessary metabolites. The metabolite production by plants cell, various characteristic features and configuration of the designed reactor contributes to the economic feasibility of the process. Though a single plant cell is used for the production of a single product, the flexible design, different flow materials with necessary aeration requirements is a need of concern. The successful exploitation of any plant cell used in suspension cell cultures will depend on multidisciplinary approaches of the plant morphogenesis.

The physical as well as chemical environment with the growth of cell biomass and controlled conditions need to be further investigated. The different chemical composition in the growth media, growth promoting regulators also need to be further studied in detail which plays a crucial role in accompanying the physical conditions. Though the efficiency of plant cell cultures is studied by few researchers, there exists few problems with the scale up processes and its relationships with different plant cells. However, the broad advantage with the use of these types of bioreactors is its high efficiency and its ease of operation. This could be of significant challenge to the process engineers for designing the bioreactors with all relevant invitro environment conditions in the bioreactors for the purpose of plant cell cultures.

Acknowledgments The authors would like to express their sincere thanks to Dr. C. Vaman Rao, HOD, Department of Biotechnology, for the necessary guidance. Due thanks are to the Principal, NMAMIT, Nitte for supporting with the department facilities.

References

Bányai, P., Bálványos, I., Kursinszki, L., Szöke, E.: Cultivation of *Lobelia inflata* L. Hairy Root culture in bioreactor. Acta Hortic. **597**, 253–256 (2003)

Behkish, A., Men, Z., Inga, R.J., Morsi, B.I.: Mass transfer characteristics in a large-scale slurry bubble column reactor with organic liquid mixtures. Chem. Eng. Sci. **57**(16), 3307–3324 (2002)

Betts, J., Baganz, F.: Miniature bioreactors: current practices and future opportunities. Microb. Cell Fact. **5**(21), 1–14 (2006)

Bouaifi, M., Hebrard, G., Bastoul, D., Roustan, M.: A comparative study of gas holdup, bubble size, interfacial area and mass transfer coefficients in stirred gas–liquid reactors and bubble columns. Chem. Eng. Process **40**(2), 97–111 (2001)

Choi, K.T., Park, J.C., Ahn, I.O.: Saponin production in tissue culture of ginseng (*Panax ginseng* C. A. Meyer). J. Ginseng Res. **14**(2), 107–111 (1990)

Deckwer, W.D., Schumpe, A.: Improved tools for bubble column reactor design and scale-up. Chem. Eng. Sci. **48**(5), 889–911 (1993)

Hyndman, C.L., Larachi, F., Guy, C.: Understanding gas-phase hydrodynamics in bubble columns: a convective model based on kinetic. Chem. Eng. Sci. **52**(1), 63–77 (1997)

Karppinen, K., Hohtola, A., Gyorgy, Z., Neubauer, P., Tolonen, A., Jalonen, J.: Comparison of growth and secondary metabolite accumulation in cultures of compact callus aggregates and shoots of *Hypericum perforatum* L. in shake flask and bubble column bioreactor. Acta Hortic. **725**, 605–612 (2006)

Klaas, V.R., Johannes, T.: Basics of Bioreactor Design, pp. 197–199. Marcel Dekker Inc., New York (1991)

Nandagopal, S., Ranjitha, K.B.D.: Phytochemical and antibacterial studies of Chicory (*C. Intybus*. L.) a multipurpose medicinal plant. Adv. Biol. Res. **1**(1–2), 17–21 (2007)

Ozlem, Y.C., Aynur, G., Fazilet, V.S.: Large Scale Cultivation of Plant Cell and Tissue Culture in Bioreactors, pp. 1–54. Transworld Research Network Kerala (2010)

Ozturk, S.S., Schumpe, A., Deckwer, W.D.: Organic liquids in a bubble column: holdups and mass transfer coefficients. Am. Inst. Chem. Eng. **33**(9), 1473–1480 (1987)

Paek, K., Chakrabarty, D., Hahn, E.: Application of bioreactor systems for large scale production of horticultural and medicinal plants. In: Liquid Culture Systems for in vitro Plant Propagation, pp. 95–116. Springer link (2005)

Pirdashti, M., Kompany, R.: Effects of height to diameter ratio and aeration rate on liquid mixing and hydrodynamic properties in a bubble column. Iran. J. Chem. Eng. **6**(3), 46–52 (2009)

Saurabh, C., Sunita, F., Ashok, K.S., Virendra, S.B.: Bioprocess considerations for production of secondary metabolites by plant cell suspension cultures. Biotechnol. Bioprocess Eng. **7**, 138–149 (2002)

Takayama, S., Misawa, M.: Mass propagation of *Begonia hiemalis* plantlets by shake culture. Plant Cell Physiol. **22**, 461–467 (1981)

Industrial Applications of Caffeine Degradation by *Pseudomonas* sp.

Swati Sucharita Dash, Sree Ahila Retnadhas, Nameeta Rao and Sathyanarayana N. Gummadi

1 Introduction

Coffee and tea are the major drinks which people consume throughout the world and they are also the rich source of caffeine, a purine alkaloid naturally present in tea leaves, coffee beans, cocoa beans, cola nuts and many other plants. Caffeine acts as a vasoconstrictor, diurant and central nervous system stimulant (Kalmar and Cafarelli 1999). Acute intake of caffeine is known to cause osteoporosis (Rapuri et al. 2001), pregnancy complications (Kuczkowski 2009) and many types of cancers (Slattery et al. 1990). Also, industries like coffee and tea processing industries which use caffeine rich starting materials release a large amount of solid and liquid wastes which are rich in caffeine. Solid wastes like coffee husks and coffee affect soil fertility and microbial growth (Friedman and Waller 1983). Caffeine in liquid effluents when released into surrounding waterbodies alters water eco-system and contaminates surface water and ground water (Glassmeyer and Shoemaker 2005). Caffeine in the effluents make the environment for biological degradation of other organic components very difficult as caffeine is toxic to majority of the microbial species. Removing caffeine from food products and industrial wastes thus becomes very vital for environmental and health safety (Dash and Gummadi 2006).

Conventional methods of caffeine removal include solvent extraction procedures where toxic solvents like ethylacetate, dichloromethane and trichloroethylene are used (Gokulakrishnan et al. 2005). In many developed countries, solvent extraction was replaced by non-toxic and safer supercritical CO_2 extractions but the process is

Swati Sucharita Dash and Sree Ahila Retnadhas have contributed equally to the work presented here.

S.S. Dash · S.A. Retnadhas · N. Rao · S.N. Gummadi (✉)
Department of Biotechnology, Bhupat and Jyoti Mehta School of Biosciences, IIT Madras, Chennai, India
e-mail: gummadi@iitm.ac.in

Prasanna B.D. et al. (eds.), *Biotechnology and Biochemical Engineering*,
DOI 10.1007/978-981-10-1920-3_19

expensive. These methods are also non-specific as they remove the flavour of food products thereby necessitating the need to add chemically synthesized flavouring compounds. Biological methods offer a cheaper and safer alternative for conventional decaffeination (Dash and Gummadi 2006). A *Pseudomonas* sp. isolated from coffee plantation soil was shown to grow on caffeine as its only source of energy, utilizing as high as 20 g/l caffeine completely within 120 h (Gummadi and Santhosh 2010). It has been reported that caffeine and its metabolites degrading enzymes are inducible (Dash and Gummadi 2008). In this study, we have used induced cells of *Pseudomonas* sp. and devised decaffeination techniques for caffeine removal from industrial effluents and black tea.

2 Materials and Methods

2.1 *Materials and Microorganism*

Effluent was obtained from Tata Coffee's Instant Coffee Division, Theni Unit, India. Liquid effluent after removing solid contents was used for effluent decaffeination experiments. AVT black tea powder was used for the black tea decaffeination experiments. *Pseudomonas* sp. used in this study was isolated in our lab and was maintained on CAS agar medium (Dash and Gummadi 2008). Induced cells were prepared by inoculating three loops full of actively growing cells in 25 ml nutrient broth (NB) and incubated at 30 °C and 180 rpm. Once A_{600} reaches 1.2–1.4, production media (CAS media) (Gummadi et al. 2009) was inoculated with 6 % seed culture (NB) and incubated at 30 °C and 180 rpm till 90–95 % of initial caffeine gets degraded. Cells were then harvested by spinning at 10,000 rpm, 4 °C for 5 min under sterile conditions.

2.2 *Degradation of Caffeine in Liquid Effluent from Instant Coffee Manufacturing Unit*

Induced cells were washed with 10 mM potassium phosphate buffer pH 7.0 and they were inoculated in the effluent so as to get different cell concentrations (0.5–8 g/l) and incubated at 30 °C and 180 rpm. Samples were collected at regular intervals for biomass growth and caffeine degradation. The same experiments were repeated in another batch with pH of the effluent adjusted to 7.8. Induced cells of *Pseudomonas* sp. were immobilized as described earlier (Gummadi et al. 2009) in various immobilization matrices like sodium alginate, agar-agar and agarose to a final concentration of 8 g/l of induced cells in the respective matrix solution. 6 and 18 % of beads (v/v) were inoculated in pH adjusted (7.8) effluent and incubated at 30 °C 180 rpm. Samples were taken at regular intervals and caffeine concentration was estimated. Re-usability of the beads was tested for 3 times.

2.3 *Degradation of Caffeine in Black Tea Powder by Induced Cells of* Pseudomonas *sp.*

Commercially available AVT black tea powder was used for all the decaffeination experiments. Induced cells of *Pseudomonas* sp. were suspended in 10 mM potassium phosphate buffer, pH 8.0 to get final concentrations of 50–600 g/l. These induced cells at various concentrations were added to black tea in the ratio 1:1 (v/w) to get concentrations of 0.05–0.6 g cells/g tea and was incubated at 30 °C for 2.5 h after proper mixing. After incubation, tea-bacteria mixture was subjected to 70 °C for 10 min to heat inactivate the bacterial cells. Caffeine and polyphenol contents were then estimated after it was dried at room temperature for 16 h. After selecting an optimum cell concentration, incubation time of tea-bacteria was optimized.

2.4 *Sequential Treatment of Black Tea Powder with Induced Cells of* Pseudomonas *sp.*

After incubation of tea-bacteria mixture at 30 °C for 2.5 h, the cells were heat inactivated at 70 °C for 10 min and then they were cooled before adding the next batch of fresh induced cells in the same 1:1 ratio. After 2.5 h incubation at 30 °C, again the process of heat inactivation and addition of fresh batch of cells were continued for 3 cycles totally. After 3 cycles, the mixture was dried at room temperature for 16 h before estimating polyphenols and caffeine content.

2.5 *Analytical Procedures*

Caffeine was estimated at 254 nm by HPLC using C-18 column as stationary phase and 30 % methanol as mobile phase. Polyphenols in black tea were estimated by following the protocol discussed by Someswararao et al. (2013). Extraction of caffeine from black tea was done following an ISO 20481 (2008) method.

3 Results and Discussion

3.1 *Degradation of Caffeine in Liquid Effluent from Instant Coffee Manufacturing Unit by Induced Cells*

Effluent was found to be acidic with pH 4.8 and initial caffeine concentration before inoculation with *Pseudomonas* sp. was 110 mg/l. At native pH, only 8 g/l of induced cells were able to degrade caffeine completely in 1 h and at concentrations less than 8 g/l (Fig. 1a). This shows the toxicity of acidic pH for the survival of *Pseudomonas* sp. However, when the pH of effluent was adjusted to 7.8 which is the optimum pH for *Pseudomonas* sp. to degrade caffeine (Dash and Gummadi 2007), even 0.5 g/l of induced cells were enough to completely degrade caffeine within half an hour. As the concentration of induced cells increases, caffeine degradation rates were also increased. At 8 g/l of induced cells, complete caffeine degradation was obtained within 30 min (Fig. 1b).

3.2 Degradation of Caffeine in Liquid Effluent from Instant Coffee Manufacturing Unit by Immobilized Cells

Different matrices were used to immobilize induced cells of *Pseudomonas* sp. Induced cells were immobilized at 8 g/l concentration in all the three matrices. Initially 6 % (v/v) of the immobilized beads were added to the effluent. It was observed that 14, 33.5 and 38 % degradation of caffeine was achieved at the end of 4 h in agar-agar, agarose and calcium alginate respectively (Fig. 1c). After 3 h when 18 % inoculum was used, 70 and 82 % degradation of caffeine was observed in agar-agar and agarose respectively, while the maximum degradation of 93 % was detected in calcium alginate (Fig. 1d). Out of three matrices used, calcium alginate was found to be more effective in retaining caffeine degrading ability of induced *Pseudomonas* sp. When 18 % inoculums was used, it was able to degrade almost 95 % of caffeine in 3 h. These calcium alginate beads were also re-usable as the efficiency in degrading caffeine was not reduced even after three runs (Fig. 1e) and no cell leaching was observed.

3.3 Degradation of Caffeine in Black Tea Powder by Induced Cells of Pseudomonas sp.

HPLC analysis of the black tea extract showed concentration of caffeine to be 122.45 mg/L. This constitutes 3.67 % dry weight of tea which correlates with the standard value of caffeine which is 3–4 % of the tea dry weight (Goto et al. 1996). Tea samples were treated with bacterial cells at concentrations ranging from 0.05 to 0.6 g cells/g tea for 2.5 h. At cell concentration of 0.05 g/g tea, the amount of caffeine removed was as low as 6 % but increased to 17 % with increase in cell concentration to 0.15 g/g tea (Fig. 2a). However, further increase in the cell loading (up to 0.6 g cells/g tea) did not enhance the amount of caffeine degraded. Hence, cell loading of 0.15 g cell/g tea was found to be optimum for caffeine removal in tea. Theaflavin level was found to be reduced at all cell concentrations when compared to untreated tea. A proportionate increase in total liquor colour was seen with increase in cell loading. At 0.15 g cell/g tea, there was no significant reduction

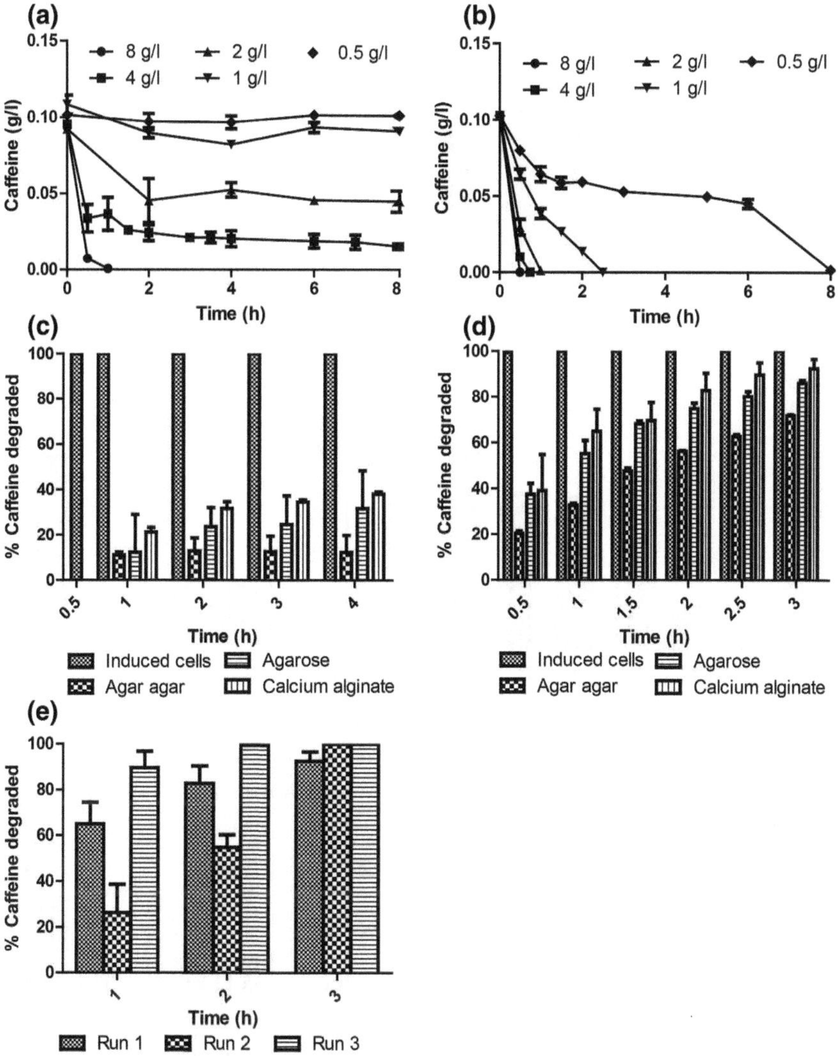

Fig. 1 Degradation of caffeine in coffee industry effluent by induced cells of *Pseudomonas* sp. Caffeine degradation profile by various concentrations of induced *Pseudomonas* sp. in effluent at **a** degradation at its native pH, **b** degradation at pH adjusted to 7.8. Degradation of caffeine in coffee industry effluent by induced *Pseudomonas* sp. immobilized in various immobilization matrices at a concentration of **c** 6 % (v/v) immobilized beads, **d** 18 % (v/v) immobilized beads, **e** re-usability of induced cells immobilized in calcium alginate beads

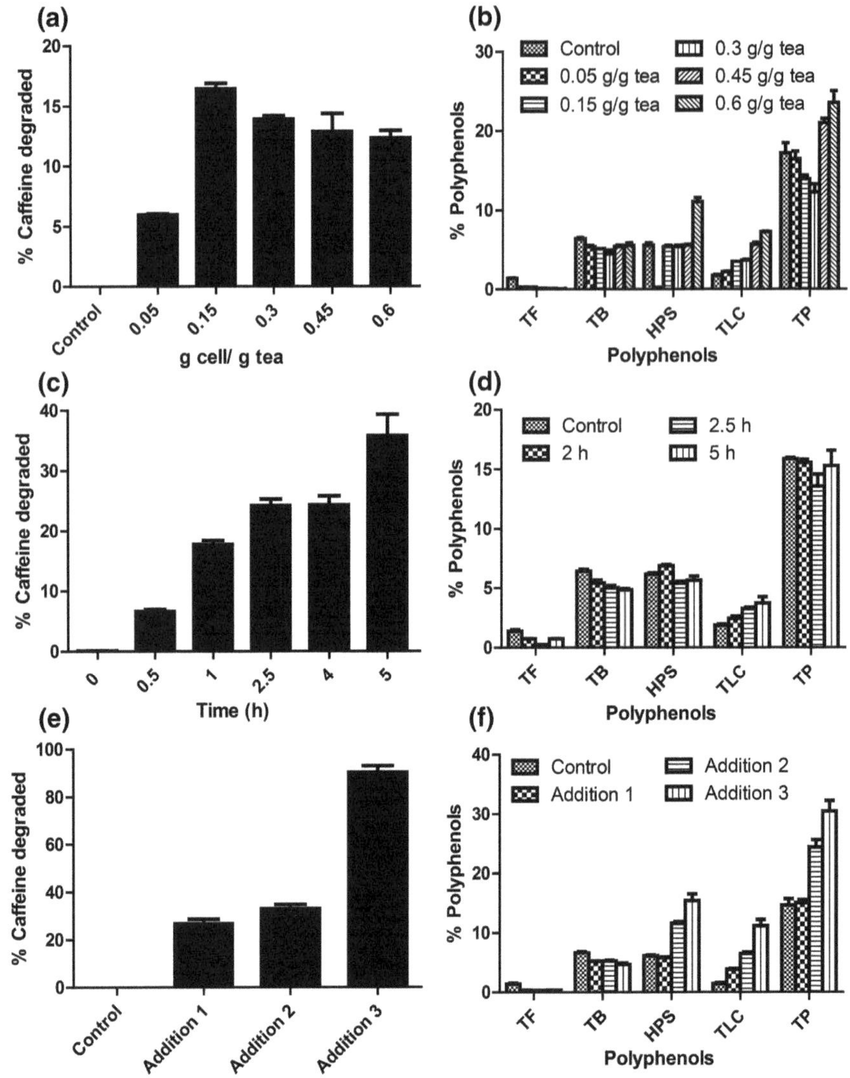

Fig. 2 Degradation of caffeine in commercially available tea samples using induced *Pseudomonas* sp. Effect of various cell concentrations on **a** caffeine content, **b** polyphenolic content of black tea samples. Effect of incubation period of induced cells with black tea on **c** caffeine content, and **d** polyphenolic content of black tea. Effect of sequential addition of induced cells to black tea samples on **e** caffeine content, and **f** polyphenolic content

in polyphenolic contents other than theaflavin (Fig. 2b). Initial cell loading was then fixed at 0.15 g cells/g tea.

Upon incubating tea-bacteria mixture for different time intervals, it was observed that the percent age of degradation increased with increase in incubation time and

22 % degradation of caffeine was achieved in 2.5 h. Maximum degradation of 33.7 % was obtained at an incubation time of 5 h (Fig. 2c). Polyphenol content analysis shows that there was no significant change in polyphenol content compared to untreated tea (Fig. 2d). It was also noted that the total liquor colour increases with time probably because of oxidation of polyphenols (Smith 2002). Since the percentage of increase in caffeine degradation from 2.5 to 5 h was not high and also to avoid more polyphenol oxidation, 2.5 h was chosen for further experiments.

3.4 *Sequential Treatment of Black Tea with Induced Cells of* **Pseudomonas** *sp.*

In an attempt to increase caffeine removal from tea, sequential bacterial treatment of tea sample was adopted. Addition of fresh cells at 0.15 g cells/g tea after 2.5 h was found to enhance caffeine degradation process. After three additions of fresh cells, 83 % of caffeine degradation was obtained (Fig. 2e). In order to detect differences between sequential bacterial treatment and one time treatment, tea samples were treated with cell concentration of 0.45 g/g tea, which added up to the amount of cells used for 3 sequential additions, and incubated at a stretch for 7.5 h. However, only 5.5 % caffeine removal was obtained on single treatment. Polyphenol analysis shows that total liquor colour and highly polymerized substances increased whereas theaflavin decreased significantly in the sequentially treated tea (Fig. 2f).

4 Conclusions

Biological decaffeination techniques can effectively overcome the disadvantages caused by conventional decaffeination techniques (Gokulakrishnan et al. 2005). This study has shown that *Pseudomonas* sp. is an efficient candidate for development of biological decaffeination techniques. Immobilization of induced cells protects them from other toxic compounds in the effluents and they are reusable thereby offer a better alternative than free cells (Aksu and Bülbül 1999). Even though induced cells were able to degrade caffeine from black tea with very less change in polyphenolic content, using whole bacterial cells to degrade caffeine in food products like tea powder is unsafe. Application of purified caffeine metabolizing enzyme from *Pseudomonas* sp. for decaffeinating food products is a better alternative which can be studied in future. Thus, this study encourages the use of safer and effective biological decaffeination techniques to be used in industries.

Acknowledgments The authors acknowledge the Department of Biotechnology (DBT), India and the Department of Science and Technology (DST), India for funding the research work described in this paper. SR also acknowledges University Grant Commission (UGC) for fellowship.

References

Aksu, Z., Bülbül, G.: Determination of the effective diffusion coefficient of phenol in Ca-alginate-immobilized *P. putida* beads. Enzyme Microb. Technol. **25**, 344–348 (1999)

Dash, S.S., Gummadi, S.N.: Catabolic pathways and biotechnological applications of microbial caffeine degradation. Biotechnol. Lett. **28**, 1993–2002 (2006)

Dash, S.S., Gummadi, S.N.: Optimization of physical parameters for biodegradation of caffeine by *Pseudomonas* sp.: a statistical approach. Am. J. Food Technol. **2**, 21–29 (2007)

Dash, S.S., Gummadi, S.N.: Inducible nature of the enzymes involved in catabolism of caffeine and related methylxanthines. J. Basic Microbiol. **48**, 227–233 (2008)

Friedman, J., Waller, G.R.: Caffeine hazards and their prevention in germinating seeds of coffee (*Coffea arabica* L.). J. Chem. Ecol. **9**, 1099–1106 (1983)

Glassmeyer, S.T., Shoemaker, J.A.: Effects of chlorination on the persistence of pharmaceuticals in the environment. Bull. Environ. Contam. Toxicol. **74**, 24–31 (2005)

Gokulakrishnan, S., Chandraraj, K., Gummadi, S.N.: Microbial and enzymatic methods for the removal of caffeine. Enzyme Microb. Technol. **37**, 225–232 (2005)

Goto, T., Yoshida, Y., Kiso, M., Nagashima, H.: Simultaneous analysis of individual catechins and caffeine in green tea. J. Chromatogr. A **749**, 295–299 (1996)

Gummadi, S.N., Santhosh, D.: Kinetics of growth and caffeine demethylase production of *Pseudomonas* sp. in bioreactor. J. Ind. Microbiol. Biotechnol. **37**, 901–908 (2010)

Gummadi, S.N., Ganesh, K.B., Santhosh, D.: Enhanced degradation of caffeine by immobilized cells of *Pseudomonas* sp. in agar-agar matrix using statistical approach. Biochem. Eng. J. **44**, 136–141 (2009)

Kalmar, J.M., Cafarelli, E.: Effects of caffeine on neuromuscular function. J. Appl. Physiol. **87**, 801–808 (1999)

Kuczkowski, K.M.: Caffeine in pregnancy. Arch. Gynecol. Obstet. **280**, 695–698 (2009)

Rapuri, P.B., Gallagher, J.C., Kinyamu, H.K., Ryschon, K.L.: Caffeine intake increases the rate of bone loss in elderly women and interacts with vitamin D receptor genotypes. Am. J. Clin. Nutr. **74**, 694–700 (2001)

Slattery, M.L., West, D.W., Robison, L.M., French, T.K., Ford, M.H., Schuman, K.L., Sorenson, A.W.: Tobacco, alcohol, coffee, and caffeine as risk factors for colon cancer in a low-risk population. Epidemiology **1**(2), 141–145 (1990)

Smith, A.: Effects of caffeine on human behavior. Food Chem. Toxicol. **40**, 1243–1255 (2002)

Someswararao, C., Srivastav, P.P., Das, H.: Quality of black teas in Indian market. Afr. J. Agric. Res. **8**, 491–494 (2013)

Highly Sensitive Determination of Ascorbic Acid, Dopamine and Uric Acid Using Mesoporous Nitrogen Containing Carbon

Anju Joshi and C.N. Tharamani

1 Introduction

Ascorbic Acid (AA), dopamine (DA), and uric Acid (UA) are physiologically essential biomolecules involved in maintenance of metabolism, central nervous system and circulation system of human body. However slight change of their concentration in extracellular fluid represent an alarming situation for a wide variety of neurodegenerative diseases, scurvy, gout, and hyperuricemia. Hence simultaneous determination of these biomolecules is of utmost importance in the field of neurochemistry, clinical diagnostics and medical applications (Mo and Ogorevc 2001).

All of these biomolecules are electroactive in nature making electrochemical determination a most convenient choice for their determination. But simultaneous determination of these biomolecules is generally hampered by the drawbacks like fouling of electrode, closely spaced oxidation potentials and homogeneous electrocatalytic oxidation of AA by the oxidation products of dopamine. Therefore suitable electrocatalyst materials are generally in great demand to modify the surface of GCE and enable highly sensitive and selective determination of these biomolecules. As compared with metal nanoparticles (Wang et al. 2012), polymers (Ensafi et al. 2010), polymers composite (Liu et al. 2007), carbonaceous material have drawn considerable attention because of their wide electrochemical window, low cost, and consistent mass scale production. As a result, more and more efforts are targeted towards development of novel carbonaceous material with tunable properties to impart novel characteristics features.

A. Joshi (✉) · C.N. Tharamani
Department of Chemistry, Indian Institute of Technology Ropar,
Rupnagar 140001, Punjab, India
e-mail: anju.joshi@iitrpr.ac.in

Prasanna B.D. et al. (eds.), *Biotechnology and Biochemical Engineering*,
DOI 10.1007/978-981-10-1920-3_20

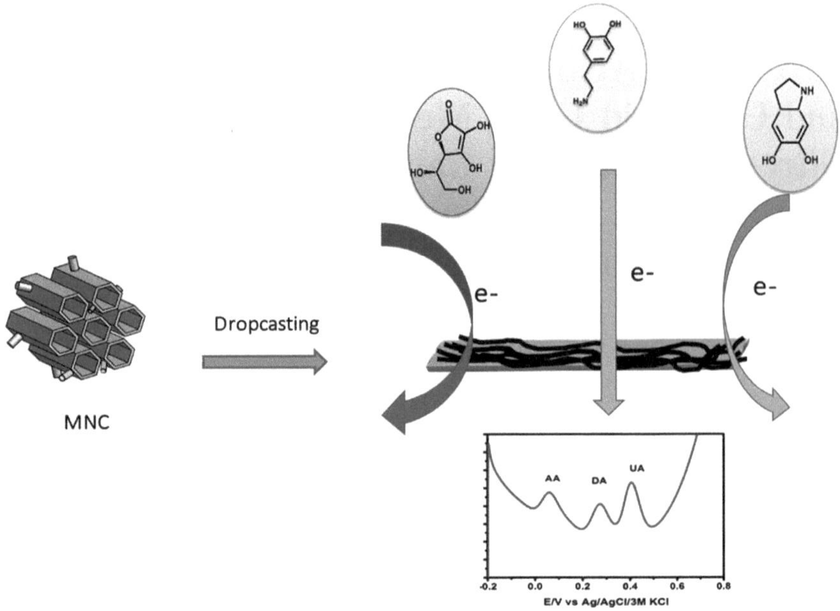

Fig. 1 Schematic representation of the electrooxidation process of *AA*, *DA* and *UA* using *MNC* modified sensor

Motivated by this fact, we here propose to synthesize novel mesoporous nitrogen containing carbon as an active catalyst material possessing graphitic structure and porous nature with multiple nitrogen containing groups enabling (Fig. 1).

2 Experimental Section

2.1 Chemicals

Ascorbic acid, dopamine hydrochloride, uric acid, Nafion and KCl. Phosphate buffer solutions (PBS 0.1 M) were prepared from stock solution of 0.1 M KH_2PO_4 and 0.1 M K_2HPO_4. All aqueous solutions were prepared with deionized water from a Millipore system (>10 MΩ cm^{-1}).

2.2 Synthesis of MNC

For synthesis of mesoporous carbon (MNC), SBA-15 was utilized as a hard template as previously reported (Vinu et al. 2007; Vinu 2008). Pyrolyzing the polymerized ethylenediamine nanocasted into a SBA-15 hard template at 600 °C results into MNC-600. The synthesized material was characterized using scanning electron microscopy (SEM, JEOL, JSM-66101 V) and XRD (PANalytical, X'Pert-Pro MPD).

2.3 Sample Preparation and Electrochemical Investigation

In order to undergo electrochemical investigation glassy carbon electrode was cleaned sequentially with alumina slurry of 3, 1, 0.3, 0.05 μm, respectively using Nylon polishing cloth (SM 407052, AKPOLISH). Since the oxidation of these biomolecules is a proton dependent process. Hence pH of the buffer solution plays an important role in driving the electrocatalytic oxidation process of AA, DA and UA. pH 6.0 was chosen on the basis of linear sweep voltammetry studies conducted at different pH buffers for further experiments. The amount of catalyst was optimized to achieve maximum sensitivity for the oxidation of AA, DA and UA by carrying out linear sweep voltammetry as shown in Fig. 2 in a solution containing 500 μM ascorbic acid with varying concentration of catalyst dropcasted on to the surface of GCE.

3 Results and Discussion

3.1 Electrochemical Determination of AA, DA and UA

Preliminary investigations were carried out to identify the potential of MNC-600 towards electrocatalytic oxidation of AA, DA and UA. For this purpose, cyclic

Fig. 2 Optimization of amount of MNC-600 catalyst dropcasted on to the surface of GCE using linear sweep voltammetry (LSV)

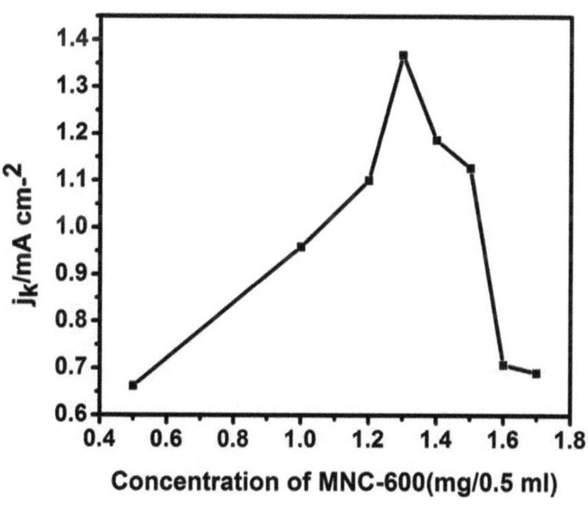

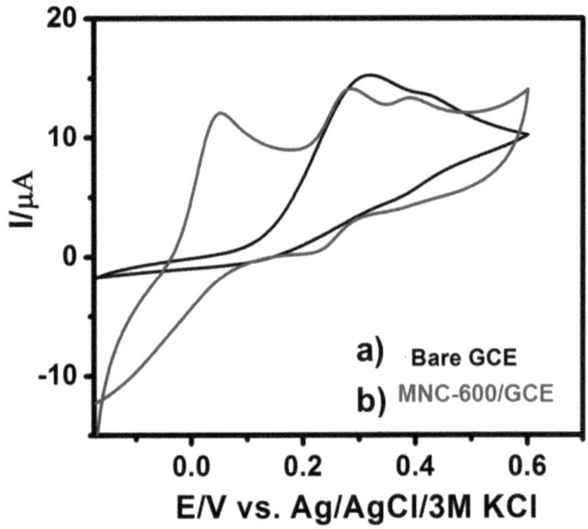

Fig. 3 Cyclic voltammetric studies at **a** bare GCE, **b** MNC-600/GCE in 0.1 M PBS (pH 6.0) solution containing 1 mM AA, 25 μM DA, 100 μM UA at a scan rate of 5 mV s^{-1}, CE: Pt wire, RE: Ag/AgCl/3M KCl

voltammetry (CV) was conducted in a mixture containing 1 mM AA, 25 μM DA and 100 μM UA at different modified surfaces (Fig. 3). CV studies shows three well separated peaks at a potentials of 49, 279 and 389 mV corresponding to electrooxidation of AA, DA and UA respectively for MNC-600/GCE which suggests that MNC-600 possess high sensitivity towards determination of AA, DA and UA with enhanced peak current and wide peak separation between oxidation potentials of analyte molecules (AA-DA and DA-UA).

Peak corresponding to AA results due to the conversion of hydroxyl group to a carbonyl group in the furan ring of AA whereas well defined redox peaks corresponds to the conversion of dopamine to dopaminoquinone (Kalimuthu et al. 2009). Similar to DA, UA also shows a well-defined oxidation peak due to the conversion of UA into quinonoid, followed by a fast electrochemical (EC) reaction (Safavi et al. 2006).

Compared to MNC-600/GCE, bare GCE shows a broadened response at a potential of 306 mV with a shoulder at about 420 mV which suggests a sluggish electron transfer kinetics (Stergiou et al. 2010) with concomitant fouling of the electrode surface due to formation of polymerized oxidation products onto the surface of GCE (Ping et al. 2012). The above mentioned remarkably high electrocatalytic activity of MNC-600 material can be ascribed to the presence of numerous nitrogen containing groups (Sheng et al. 2012) and highly porous nature of MNC material.

To gain insights into the mechanism towards electro-oxidation process, cyclic voltammetric studies were conducted at varying scan rates (5–100 mV/s) for individual biomolecule. A continuos increase in peak current with increase in scan rate is observable for all biomolecule as shown in Fig. 4 which suggests the electrooxidation process to be diffusion controlled (Yuan et al. 2012).

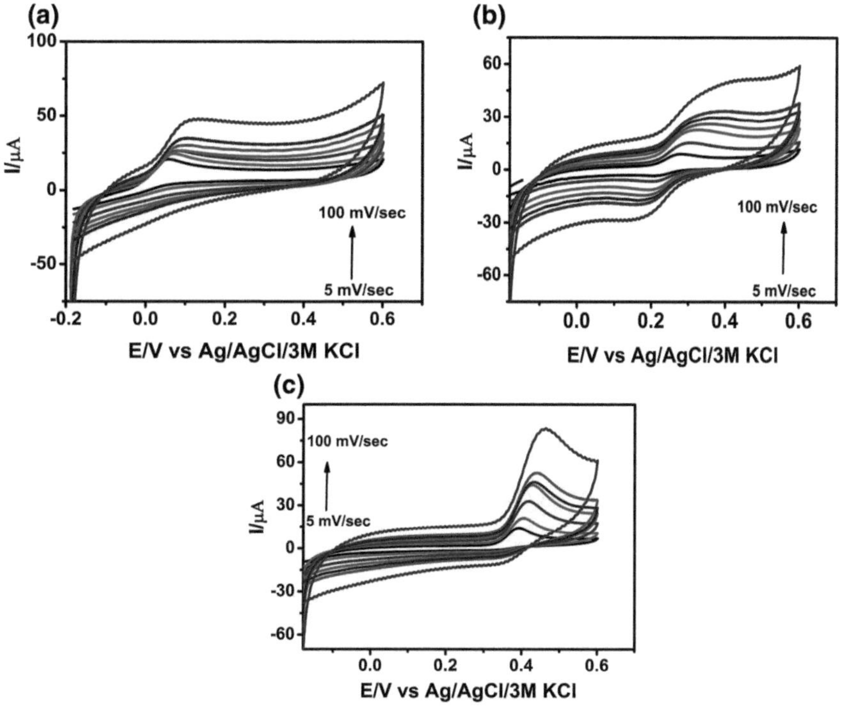

Fig. 4 Cyclic voltammograms of a MNC-600 modified GCE in 0.1 M PBS (pH 6.0) solution containing **a** 500 μM, **b** 25 μM, **c** 40 μM UA at various scan rates ranging from 5 to 100 mV s^{-1}

To explore the potential of MNC-600 towards sensitive and selective determination of AA, DA and UA linear sweep voltammetry was employed at variable concentration of concerned biomolecule while keeping the concentration of other two biomolecules at constant. It is quite clear from Fig. 5 that there is a sequential increase in peak current with subsequent increase in concentration of the concerned biomolecule while peak current of other two biomolecule remains constant.

The linear range for simultaneous determination was evaluated out to be 0.01–500 μM, 0.001–50 μM, 0.01–200 μM. Both MNC material possess excellent sensitivity towards simultaneous determination of ascorbic acid, dopamine and uric

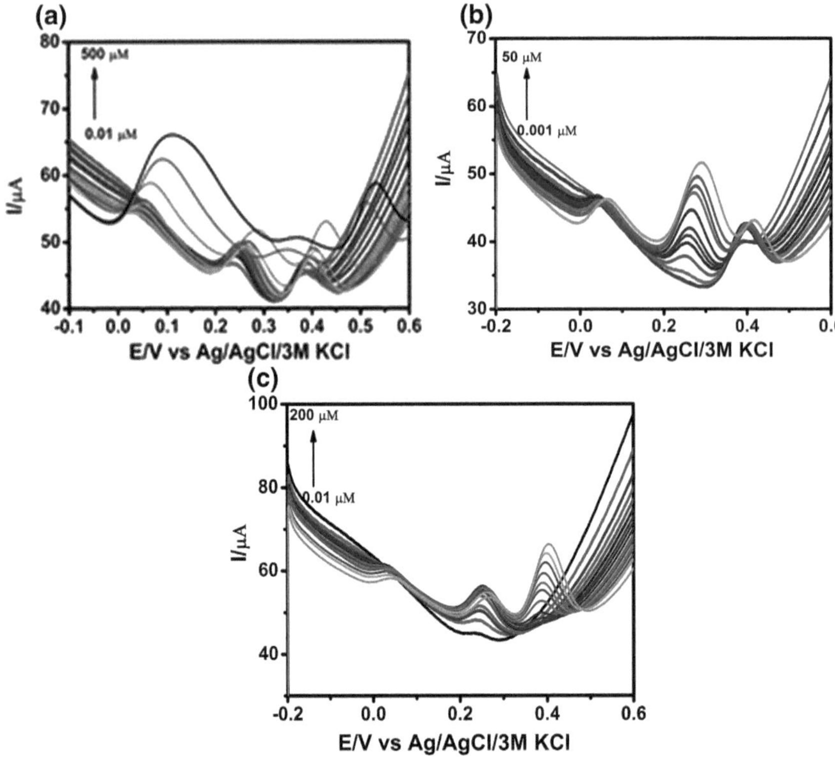

Fig. 5 Linear sweep voltammograms over MNC-600 in 0.1 M PBS (pH 6.0) solution containing (**a**) 1 μM DA and 1 μM UA and different concentrations of AA ranging from 0.01 to 500 μM, (**b**) 100 μM AA and 1 μM UA and different concentrations of DA ranging from 0.001 to 50 μM, (**c**) 1 μM DA and 100 μM AA and different concentrations of UA ranging from 0.01 to 200 μM at a frequency of 15 Hz, step potential 0.3 mV, pulse amplitude 25 mV, scan rate 5 mV s^{-1}, CE: Pt wire, RE: Ag/AgCl/3M KCl

acid. Moreover, peak separation remains sufficient to study the analyte of interest in presence of coexisting biomolecules. Furthermore, lowest detection limit and detection range obtained for the simultaneous determination of AA, DA and UA using MNC-600/GCE has been compared with existing literature which is mentioned in Table 1. It is quite clear from Table 1 that MNC-600 possess high sensitivity towards simultaneous determination of AA, DA and UA.

Table 1 Comparison of the electrocatalytic activity for the simultaneous detection of biomolecules over different electrodes

Electrode material	Linear range (μM)			Detection limit (μM)			Reference
	AA	DA	UA	AA	DA	UA	
Palladium nanoparticle-loaded carbon nanofibers modified carbon paste electrode	50–4000	0.5–160	2–200	15	0.2	0.7	Huang et al. (2008)
N-doped graphene	5–1300	0.5–170	0.1–20	2.2	0.25	0.045	Sheng et al. (2012)
Multiwall carbon nanotubes (MWCNT) modified carbon-ceramic electrode	15–800	0.5–100	0.55–90	7.71	0.31	0.42	Habibi et al. (2010)
Mesoporous carbon nanofiber-modified pyrolytic graphite electrode	100–10,000	0.05–30	0.5–120	50	0.02	0.2	Yue et al. (2012)
Mesoporous nitrogen rich carbon/GCE	1–2000	0.001–400	0.01–1500	1	0.001	0.01	This work

4 Conclusion

MNC-600 material possess excellent sensitivity towards determination of ascorbic acid, dopamine and uric acid due to their high surface area and N-content in the carbon material. MNC-600 modified electrodes show well separated potentials for the oxidation of AA, DA and UA, suggesting the applicability of these materials for simultaneous determination of these analytes with enhanced selectivity and sensitivity. These favourable features depict promising ability of MNC catalyst modified electrodes for the simultaneous determination of AA, DA and UA.

References

Ensafi, A.A., Taei, M., Khayamian, T.: Simultaneous determination of ascorbic acid, dopamine, and uric acid by differential pulse voltammetry using tiron modified glassy carbon electrode. Int. J. Electrochem. Sci. **5**, 116–130 (2010)

Habibi, B., Pournaghi-Azar, M.H.: Simultaneous determination of ascorbic acid, dopamine and uric acid by use of a MWCNT modified carbon-ceramic electrode and differential pulse voltammetry. Electrochim. Acta **55**, 5492–5498 (2010)

Huang, J., Liu, Y., Hou, H., You, T.: Simultaneous electrochemical determination of dopamine, uric acid and ascorbic acid using palladium nanoparticle-loaded carbon nanofibers modified electrode. Biosens. Bioelectron. **24**, 632–637 (2008)

Kalimuthu, P., John, S.A.: Modification of electrodes with nanostructured functionalized thiadiazole polymer film and its application to the determination of ascorbic acid. Electrochim. Acta **55**, 183–189 (2009)

Liu, A., Honma, I., Zhou, H.: Simultaneous voltammetric detection of dopamine and uric acid at their physiological level in the presence of ascorbic acid using poly (acrylic acid)-multiwalled carbon-nanotube composite-covered glassy-carbon electrode. Biosens. Bioelectron. **23**(1), 74–80 (2007)

Mo, J.-W., Ogorevc, B.: Simultaneous measurement of dopamine and ascorbate at their physiological levels using voltammetric microprobe based on overoxidized poly (1,2-phenylenediamine)-coated carbon fiber. Anal. Chem. **73**, 1196–1202 (2001)

Ping, J., Wu, J., Wang, Y., Ying, Y.: Simultaneous determination of ascorbic acid, dopamine and uric acid using high-performance screen-printed graphene electrode. Biosens. Bioelectron. **34**, 70–76 (2012)

Safavi, A., Maleki, N., Moradlou, O., Tajabadi, F.: Simultaneous determination of dopamine, ascorbic acid, and uric acid using carbon ionic liquid electrode. Anal. Biochem. **359**, 224–229 (2006)

Sheng, Z.H., Zheng, X.Q., Xu, J.Y., Bao, W.J., Wang, F.B., Xia, X.H.: Electrochemical sensor based on nitrogen doped graphene: simultaneous determination of ascorbic acid, dopamine and uric acid. Biosens. Bioelectron. **34**(1), 125–131 (2012)

Stergiou, D.V., Diamanti, E.K., Gournis, D., Prodromidis, M.I.: Comparative study of different types of graphenes as electrocatalysts for ascorbic acid. Electrochem. Commun. **12**, 1307–1309 (2010)

Vinu, A.: Two-dimensional hexagonally-ordered mesoporous carbon nitrides with tunable pore diameter, surface area and nitrogen content. Adv. Funct. Mater. **18**(5), 816–827 (2008)

Vinu, A., Srinivasu, P., Sawant, D.P., Mori, T., Ariga, K., Chang, J.S., Hwang, Y.K.: Three-dimensional cage type mesoporous CN-based hybrid material with very high surface area and pore volume. Chem. Mater. **19**(17), 4367–4372 (2007)

Wang, C., Yuan, R., Chai, Y., Chen, S., Hu, F., Zhang, M.: Simultaneous determination of ascorbic acid, dopamine, uric acid and tryptophan on gold nanoparticles/overoxidized-polyimidazole composite modified glassy carbon electrode. Anal. Chim. Acta **741**, 15–20 (2012)

Yuan, D., Yuan, X., Zhou, S., Zou, W., Zhou, T.: N-Doped carbon nanorods as ultrasensitive electrochemical sensors for the determination of dopamine. RSC Adv. **2**, 8157–8163 (2012)

Yue, Y., Hu, G., Zheng, M., Guo, Y., Cao, J., Shao, S.: A mesoporous carbon nanofiber-modified pyrolytic graphite electrode used for the simultaneous determination of dopamine, uric acid, and ascorbic acid. Carbon **50**, 107–114 (2012)

Nano-aptamer Based Quantitative Detection of Chloramphenicol

Richa Sharma, K.V. Ragavan, K.S.M.S. Raghavarao and M.S. Thakur

1 Introduction

Chloramphenicol is a highly-effective antibiotic used widely in the treatment of animals. However, prolonged intake of CAP is associated with aplastic anaemia (AA) and bone marrow suppression in humans with percent mortality of 40–50 % (Shukla et al. 2011). Most countries, including India, have prohibited usage of CAP in production of animal food (Food Safety and Standards Regulations 2011) while maximal permissible levels is 0.3 parts per billion (ppb) in milk (Vora and Raikwar 2013). However, in recent years traces of it have been detected in milk, honey, shrimp and other aquaculture products raising serious concerns about health of consumers and the export of antibiotic-containing honey from India (Shukla et al. 2011; Islam et al. 2014).

Most available assays are based on antibodies employing conjugated enzymes. These assays, though sensitive, suffer from issues such as adjuvant addition, high cost, cumbersome and time-consuming primary and secondary antibody generation (Samsonova et al. 2012). Thus, need exists for quick, simple and cost-effective yet sensitive detection of CAP.

The proposed assay is based on salt-induced aggregation of gold nanocolloids that leads to red-to-blue colour change. Gold nanocolloids (5–100 nm) show

R. Sharma · K.V. Ragavan · K.S.M.S.Raghavarao (✉)
Academy of Scientific and Innovative Research (AcSIR), CSIR-Central Food Technological Research Institute (CSIR-CFTRI), Mysore, India
e-mail: raghavarao@cftri.res.in

R. Sharma · K.V. Ragavan · K.S.M.S.Raghavarao
Department of Food Engineering, CSIR-Central Food Technological Research Institute (CSIR-CFTRI), Mysore, India

M.S. Thakur (✉)
Materials Science Centre, University of Mysore, Mysore, India
e-mail: msthakur@yahoo.com

Prasanna B.D. et al. (eds.), *Biotechnology and Biochemical Engineering*,
DOI 10.1007/978-981-10-1920-3_21

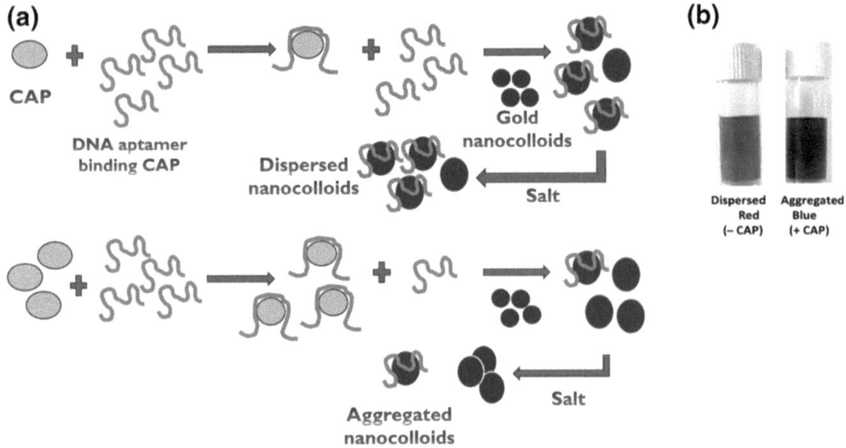

Fig. 1 **a** Principle of gold nanocolloid and aptamer-based colorimetric determination of CAP. **b** Dispersed and aggregated nanocolloids

remarkable size-dependent optical properties. They are being widely used in detection systems, a major portion of which constitutes of colorimetric aggregation based bioassays (Katiyar et al. 2013; Sharma et al. 2015b). In this technique dispersed gold nanocolloids (usually red in colour) are protected by a biorecognition element from salt-induced aggregation, the absence of which leads to the formation of larger particles changing the colour to bluish purple. When added, analyte binds and engages the biorecognition element that can no more protect the nanocolloids. The salt ions interfere with the electrostatic repulsive forces between nanoparticles, leading to formation of bigger clumps. The resulting colour change is observed visually and measured spectrophotometrically.

Aptamers are highly specific biorecognition elements that change conformation on analyte binding (Sharma et al. 2015b). The present method involves protection of nanocolloids by a DNA aptamer that has a strong affinity for CAP, to produce a fast and quantitative response in its detection in buffer (principle explained in Fig. 1). Gold nanocolloid based colorimetric aptasensing has been used to detect antibiotics in previous studies (Kim et al. 2010; Derbyshire et al. 2012).

2 Materials and Methods

2.1 Materials

Tetrachloroauric acid trihydrate, trisodium citrate, Tris-HCl and aptamer for CAP (5′AGCAGCACAGAGGTCAGATGACTTCAGTGAGTTGTCCCACGGTCGGC GAGTCGGTGGTAGCCTATGCGTGCTACCGTGAA-3′) were obtained from Sigma-Aldrich, USA. All other chemicals were procured from Ranbaxy Fine

Chemicals Limited, India. All reagents used were extra pure and of analytical grade. Three-stage MilliQ-Millipore plus purification system was used for purification of water used in the experiments. Buffer used was as reported (Mehta et al. 2011). Aptamer stock was prepared in water and stored at −20 °C.

2.2 *Methods*

2.2.1 Synthesis and Characterization of Gold Nanocolloids

Citrate-capped gold nanocolloids were synthesized according to our previously reported protocol (Sharma et al. 2015a). Size and concentration of the nanocolloids were calculated using the absorbance data (Haiss et al. 2007) and confirmed by Transmission Electron Microscopy (TEM) (Jeol 2100, USA).

2.2.2 Optimization of Parameters

The following parameters and reaction conditions were optimized: binding buffer for optimal binding of aptamer with CAP (5–100 % concentration of the reported buffer composition), NaCl concentration for the least salt amount required for nanoparticle aggregation (0.05–0.6 M at intervals of 0.05 M), incubation time for the appropriate interaction of ions and aptamers with nanocolloids and analyte (5, 10, 15, 20 and 25 min) and aptamer concentration for the minimum quantity required to protect the nanocolloids from aggregation (0.2–4 μM). The added volumes are mentioned in Table 1.

2.2.3 Detection of CAP

With optimized conditions, the assay was developed for CAP detection. Different concentrations of CAP was dissolved in buffer and 25 μL of the solution was added

Table 1 Optimization of parameters and corresponding reaction composition

Parameter optimized	Composition of reaction mixture (Volumes in μL)					
	Gold nanocolloids	Binding buffer	NaCl	Aptamer	H_2O	Total volume
Binding buffer concentration	300	75[a]	–	–	75	450
NaCl concentration		75[b]	50[a]	–	25	
Incubation time		75[b]	50[b]	–	25	
Aptamer concentration		75[b]	50[b]	25[a]	–	

[a]Different concentrations
[b]Optimized concentration

to the reaction mixture comprising of 300 μL gold nanocolloids and optimized concentrations of 50 μL buffer, 25 μL aptamer and 50 μL NaCl. The extinction spectra were recorded (Spectrophotometer UV-1601, Shimadzu, Japan) and the least concentration that could be spectrophotometrically determined was noted.

3 Results and Discussion

3.1 Synthesis and Characterization of Gold Nanocolloids

The UV-Vis spectra of the nanocolloids show that the surface plasmon resonance (SPR) peak occurs at 520 nm, where the absorbance for as-prepared nanocolloids was 0.518 (Fig. 2a). According to Haiss et al. (2007), the calculated size was 7 nm and concentration was found to be 1.83×10^{-8} M. The TEM images conform to the calculated size (Fig. 2b).

3.2 Optimization of Parameters

Nanocolloid aggregation led to the formation of larger particles that corresponded to a second extinction peak. For quantitative estimation of the aggregation the ratio of the absorbance at 610 nm (A_{610}, where secondary peak begins to form) to that at the SPR wavelength (in our case 520 nm, A_{520}) was calculated and plotted against the parameter to be measured (Derbyshire et al. 2012). The same has been com-

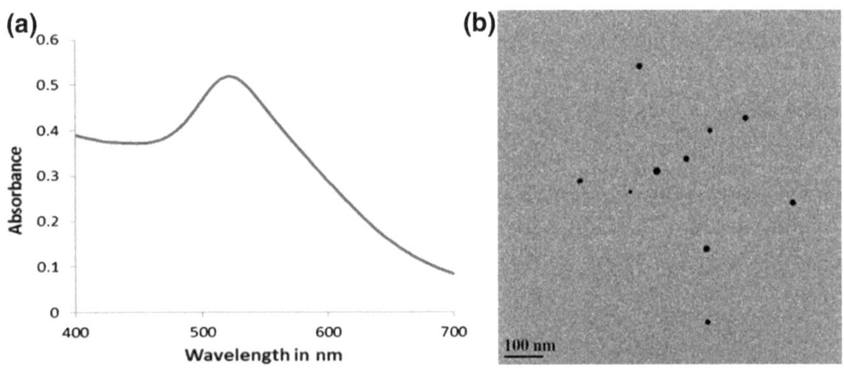

Fig. 2 **a** UV-Visible extinction spectrum of as-prepared gold nanocolloids. **b** Transmission electron micrograph image of the as-prepared nanocolloids revealing particle size of around 7 nm

pared with control values for clarification of the obtained results.

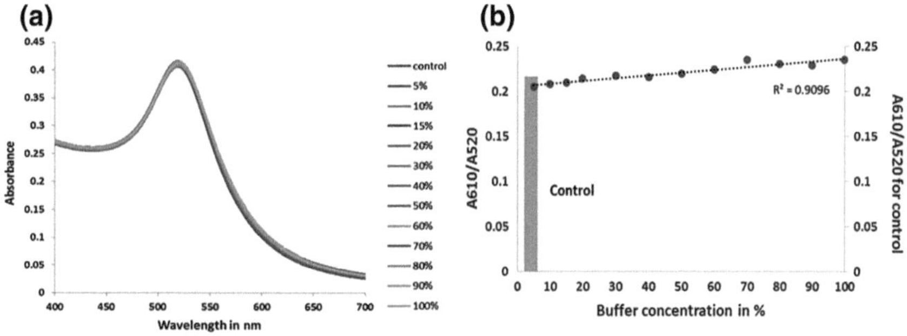

Fig. 3 **a** Extinction spectra showing effect of different concentrations of buffer on nanocolloids. Spectral lines overlap. **b** A_{610}/A_{520} values plotted against different buffer concentration

3.2.1 Buffer

Variation in concentration of buffer was not found to affect aggregation with respect to control, as seen from Fig. 3a, b. The spectral lines overlap, and the A_{610}/A_{520} values are almost in the same range as the control (Fig. 3b). As such the buffer without any dilution (100 %) was used in the assay.

3.2.2 NaCl Concentration

Gradual increase in NaCl concentration was found to aggregate more nanocolloids (Fig. 4a). It was seen that after 0.4 M of NaCl there was no significant increase in aggregation (Fig. 4b), and thus this concentration was chosen for further experiments.

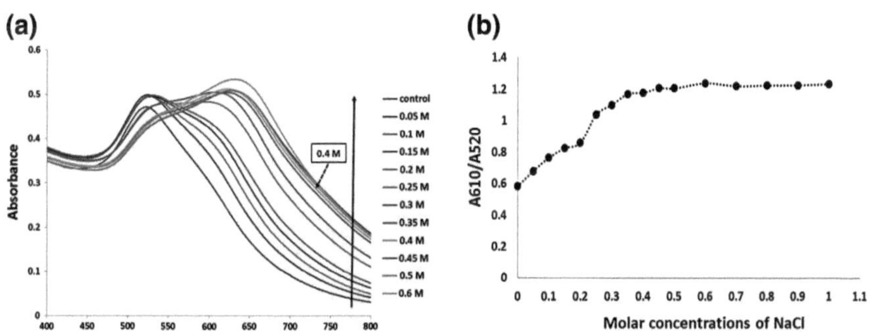

Fig. 4 **a** Aggregation pattern of gold nanocolloids using different concentrations of NaCl. **b** A_{610}/A_{520} values plotted against different NaCl concentrations. Control is 0 M NaCl

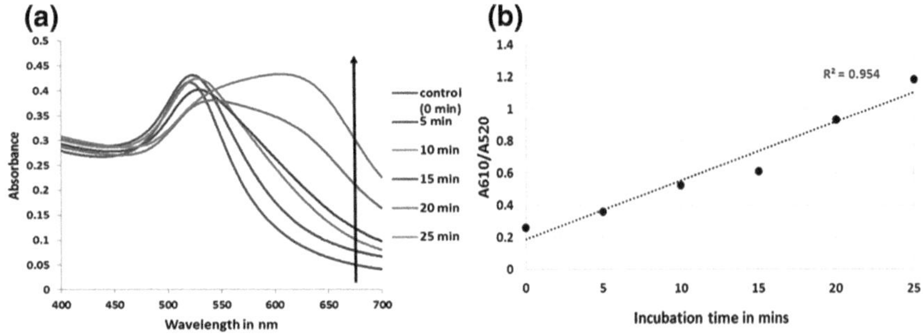

Fig. 5 **a** Effect of incubation time on aggregation of gold nanocolloids. **b** Linear increase in aggregation with increasing incubation time

3.2.3 Incubation Time

It was found that after 25 min of incubation of nanoparticles with 0.4 M NaCl complete aggregation could be achieved (Fig. 5a, b). For all later assays, this time period was used.

3.2.4 Aptamer Concentration

As aptamer concentration was increased, fewer nanocolloids were aggregated. Control I consisted of gold nanocolloids, buffer and salt i.e. complete aggregation. Control II consisted of only gold nanocolloids and buffer i.e. absence of aggregation. From Fig. 6a, b, it can be seen that 0.8 μM was able to protect nanocolloids i.e. it is closer to A_{610}/A_{520} value of Control II. It can be noted that even higher amount of aptamers can protect nanoparticles, but more number of aptamer molecules may prevent aggregation of nanoparticles even after binding to analyte. As such minimum quantity should be used. Thus, 0.8 μM aptamer concentration was chosen for further experiments.

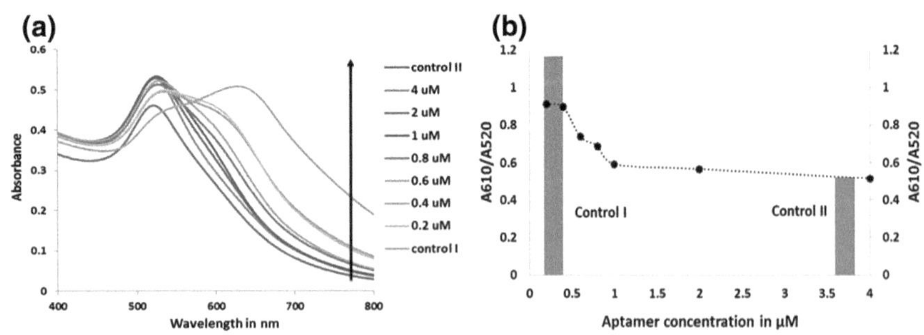

Fig. 6 **a** Increased protection of nanocolloids from aggregation by increasing aptamer concentration. **b** A_{610}/A_{520} values plotted against aptamer concentration, compared with control values

3.3 Estimation of CAP

Following the hypothesis and using optimized values of the reaction components, the assay to estimate CAP was performed. Concentrations of CAP differing by 100 units could be resolved by the detection system from 0.55 pg/mL to 55 µg/mL within which linear increase was observed (Fig. 7a, b). The lowest limit that could be reached using this technique was 0.55 pg/mL of CAP (final concentration in solution).

As is evident from the above data, the values of A_{610}/A_{520} are quite close, which can be attributed to the relatively large size of the aptamer (80 bases) and the small size of the analyte (molecular weight 323 g/mol). Some part of the aptamer may be protecting the colloid nanoparticles even after binding to CAP. In such cases, even the presence of CAP may show very less aggregation (Fig. 8). It has been reported that development and application of small molecule aptamers is challenging

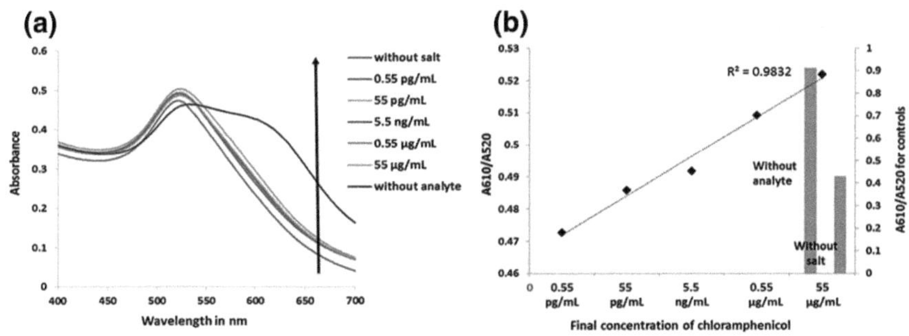

Fig. 7 **a** Absorption curves of nanocolloids in presence of different concentrations of CAP. **b** Linear quantification of CAP using the proposed system (compared to controls without aggregation and with complete aggregation)

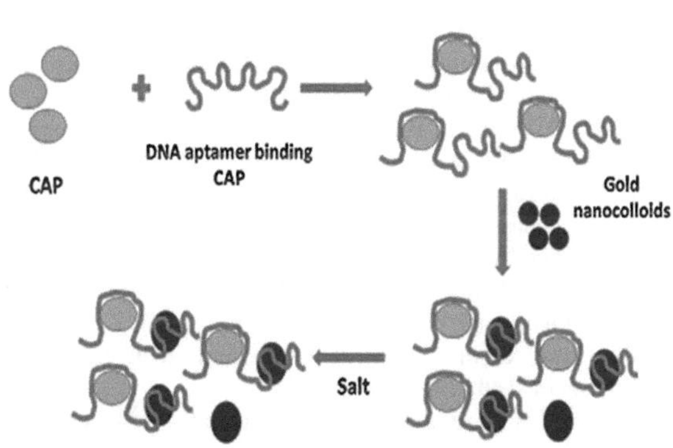

Fig. 8 Probable explanation for less A_{610}/A_{520} values in CAP detection. Even after aptamer binds CAP, it protects nanocolloids and fails to produce significant colour change on salt addition

(McKeague and DeRosa 2012). Often there is no sufficient change in conformation of the aptamers and in such cases aptamers need to be hybridized with a nucleotide or split into parts. Even for such modifications, the structure of the aptamer should be clearly elucidated, however, such information is scanty. As observed in the present detection system, colorimetric aptasensing in unmodified form faces challenges concerning sensitivity in real samples (Kim and Gu 2014). The technique has been used successfully for certain small molecules, but in those cases the aptamer length was much shorter as compared to the present system (Katiyar et al. 2013).

4 Conclusion

In the present work, we report a simple, convenient and cost-effective nanocolloid and aptamer-based bioanalytical system for chloramphenicol in buffer. The assay time is 30–35 min. The cost of around 10 assays is less than Rs. 20. Only instrumentation involved is a spectrophotometer. The detection limit is 0.55 pg/mL that is equivalent to 0.55 parts per trillion, a value much lower than the acceptable limit of CAP in food (0.3 ppb). This biosensing technique can be used for CAP detection in food and medical samples, aiding regulatory bodies and health professionals.

Acknowledgments The authors are grateful to the Director, CFTRI for providing necessary laboratory facilities. R.S. and K.V.R thank CSIR for providing research fellowship.

References

Derbyshire, N., White, S.J., Bunka, D.H., Song, L., Stead, S., Tarbin, J., Sharman, M., Zhou, D., Stockley, P.G.: Toggled RNA aptamers against aminoglycosides allowing facile detection of antibiotics using gold nanoparticle assays. Anal. Chem. **84**, 6595–6602 (2012)

Food Safety and Standards (Contaminants, Toxins and Residues) Regulations: Food Safety and Standards Authority of India, FNo 2-15015/30/2010. Accessed 21 Feb 2015 (2011)

Haiss, W., Thanh, N.T.K., Aveyard, J., Fernig, D.G.: Determination of size and concentration of gold nanocolloids from UV-Vis spectra. Anal. Chem. **79**, 4215–4221 (2007)

Islam, M.J., Liza, A.A., Reza, A.H.M.M., Reza, M.S., Khan, M.N.A., Kamal, M.: Source identification and entry pathways of banned antibiotics nitrofuran and chloramphenicol in shrimp value chain of Bangladesh. Eurasia J. Biosci. **8**, 71–83 (2014)

Katiyar, N., Selvakumar, L.S., Patra, S., Thakur, M.S.: Gold nanoparticles based colorimetric aptasensor for theophylline. Anal. Methods **5**, 653–659 (2013)

Kim, Y.S., Gu, M.B.: Advances in aptamer screening and small molecule aptasensors. Adv. Biochem. Eng. Biotechnol. **140**, 29–67 (2014)

Kim, Y.S., Kim, J.H., Kim, I.A., Lee, S.J., Jurng, J., Gu, M.B.: A novel colorimetric aptasensor using gold nanoparticle for a highly sensitive and specific detection of oxytetracycline. Biosens. Bioelectron. **26**, 1644–1649 (2010)

McKeague, M., DeRosa, M.C.: Challenges and opportunities for small molecule aptamer development. J. Nucleic Acids. Article ID 748913, 20 pages (2012)

Mehta, J., Dorst, B.V., Rouah-Martina, E., Herrebout, W., Scippo, M.-L., Blust, R., Robbens, J.: In vitro selection and characterization of DNA aptamers recognizing chloramphenicol. J. Biotech. **155**, 361–369 (2011)

Samsonova, J.V., Cannavan, A., Elliot, C.T.: A critical review of screening methods for the detection of chloramphenicol, thiamphenicol and florfenicol residues in food-stuffs. Crit. Rev. Anal. Chem. **42**, 50–78 (2012)

Sharma, R., Ragavan, K.V., Abhijith, K.S., Akanksha, Thakur M.S.: Synergistic catalysis by gold nanoparticles and metal ions for enhanced chemiluminescence. RSC Adv. **5**, 31434–31438 (2015a)

Sharma, R., Ragavan, K.V., Thakur, M.S., Raghavarao, K.S.M.S.: Recent advances in nanoparticle based aptasensors for food contaminants. Biosens. Bioelectron. **74**, 612–627 (2015b)

Shukla, P., Bansode, F.W., Singh, R.K.: Chloramphenicol toxicity: a review. J. Med. Med. Sci. **2**, 1313–1316 (2011)

Vora, V.R., Raikwar, M.K.: Determination of chloramphenicol and thiamphenicol residues in fish, shrimp and milk by ESI-LCMSMS. Int. J. Agr. Food Sci. Tech. **4**, 823–828 (2013)

Influence of Substrate Concentration, Nutrients and Temperature on the Biodegradation of Toluene in a Differential Biofilter Reactor

Suganya Baskaran, Shri Vaishnavi Perumal Selvakumar, Roshni Mohan, Rhea Mariam John, Swaminathan Detchanamurthy, Meyyappan Narayanan and Peter Alan Gostomski

1 Introduction

Biofiltration is a combination of different physical and biological processes: absorption, adsorption and degradation of the gaseous contaminant and desorption of the degradation products (Vergara-Fernández et al. 2007). Biofilter performance can be directly influenced by various operational parameters such as the filter bed type, water content, nutrient supplies, contaminant concentrations, temperature and gas flow rates (Abdel-El-Haleem 2004; Jorio et al. 2000). Understanding the impact of these environmental parameters in a biofilter is very important both at industrial scale and in lab scale biofilters. The differential biofiltration reactor system is an effective tool for investigating these operational parameters on the pollutant degradation rate. This article will discuss the outcome of the studies with a lab scale differential biofiltration reactor to understand the impact of substrate/pollutant concentration, nutrients and temperature on the toluene elimination capacity (EC) in soil.

S. Baskaran · S.V.P. Selvakumar · R. Mohan · R.M. John · S. Detchanamurthy (✉) · M. Narayanan
Department of Chemical Engineering, Sri Venkateswara College of Engineering, Pennalur, Sriperumbudur 602117, Tamil Nadu, India
e-mail: dswami@svce.ac.in

P.A. Gostomski
Department of Chemical and Process Engineering, University of Canterbury, Christchurch 8041, New Zealand

Prasanna B.D. et al. (eds.), *Biotechnology and Biochemical Engineering*,
DOI 10.1007/978-981-10-1920-3_22

Pollutant/Substrate Concentration

Knowledge about the effect of pollutant concentration on degradation is essential. In a biofilter, elimination capacity (EC) is used to describe the degradation rate normalised by bed volume (g m^{-3} h^{-1}). The microbial degradation kinetics in a biofilter, as a function of pollutant concentration, are usually described using a modified form of Monod's growth equation (Eq. 1) (Doran 1995). Degradation in a biofilm is actually a combination of microbial kinetics and mass transfer through the biofilm, but without explicit knowledge of the biofilm area and thickness, this lumped approach is commonly used. When substrate inhibition affects the degradation rate, Andrew's substrate inhibition equation (Andrews 1968) is one option that can be used (Eq. 2).

$$EC = \frac{\mathrm{EC}_{\max} S}{K_S + S} \tag{1}$$

$$\mathrm{EC} = \frac{\mathrm{EC}_{\max}\mathrm{S}}{\mathrm{K_S} + \mathrm{S} + \mathrm{S}^2/\mathrm{K_i}} \tag{2}$$

where, EC is elimination capacity in g m^{-3} h^{-1}, EC_{max} is maximum elimination capacity, S is substrate concentration in g m^{-3}, K_i is substrate inhibition constant in g m^{-3} and K_s is substrate half saturation constant in g m^{-3}. Equation 2 does not have mass transfer as an explicit term. However as K_s and K_i are lumped parameters, substrate concentration can influence EC directly through microbial kinetics or indirectly through a higher mass transfer rate causing more biofilm to be engaged in the degradation process.

Nutrients

Microorganisms present in the biofilter bed use the carbon present in the pollutant [e.g. volatile organic compound (VOC)] typically as an energy source and potentially as a carbon source for cell material, if the nutrient balance in the bed material allows for growth, as carbon is the most important building block in any cell (Kennes and Veiga 2001). After carbon, nitrogen is the next most essential nutrient for microbial growth. It makes up about 15 % of the dry cell mass and is a major constituent of nucleic acids and proteins. Next to nitrogen, phosphorus, sulphur and potassium are considered essential for many intracellular processes in a microorganism (ATP production, disulphate bond formation, maintenance of cellular pH, etc.) (Alahari and Apte 2004; du Plessis et al. 1998; Nikiema et al. 2010; Sorial et al. 1998). Therefore, the concentration, frequency and type of nutrients needed for treating different gaseous pollutants with various biofilter bed media remains highly empirical.

Temperature

Environmental conditions can raise or lower the bed temperature through solar input, winter cooling or upstream process changes. The optimum temperature range for toluene degraders is usually reported as 30–40 °C (Kiared et al. 1997; Leson and Winer 1991; Ottengraf and Vandenoever 1983). However, based on the type of microorganisms involved in degrading the particular pollutant, the optimum temperature range will change for attaining the maximum degradation rate. Hence, the major objective of this article is to understand the effects of toluene concentration, different nutrients and temperature on the toluene elimination capacity using soil with a differential biofilter with excellent control of the water environment. These studies will help further to identify a suitable toluene concentration range and working temperature to attain high EC.

2 Materials and Methods

2.1 Differential Biofilter Reactor

The design of the differential biofiltration reactor used in the current study exposes the whole biofilter bed medium to the same environmental conditions and residual pollutant concentration (Fig. 1). This is in contradictory to a typical column biofilter where most of the parameters change along the length of the reactor. A matric potential of −10 cm H_2O is maintained throughout this study as previous work demonstrated this was in the optimum water content range. The desired inlet toluene concentration is generated by a diffusion flask-water bath set up (Detchanamurthy and Gostomski 2012) and the air flow rate is controlled by a mass flow controller (M100B, MKS Instruments, Andover, MA, USA). Before the toluene-laden air enters the reactor, it is humidified by a shell and tube humidifier (Perma Pure LLC, Toms River, NJ, USA). Soil (Park house Garden Supplies, C/N

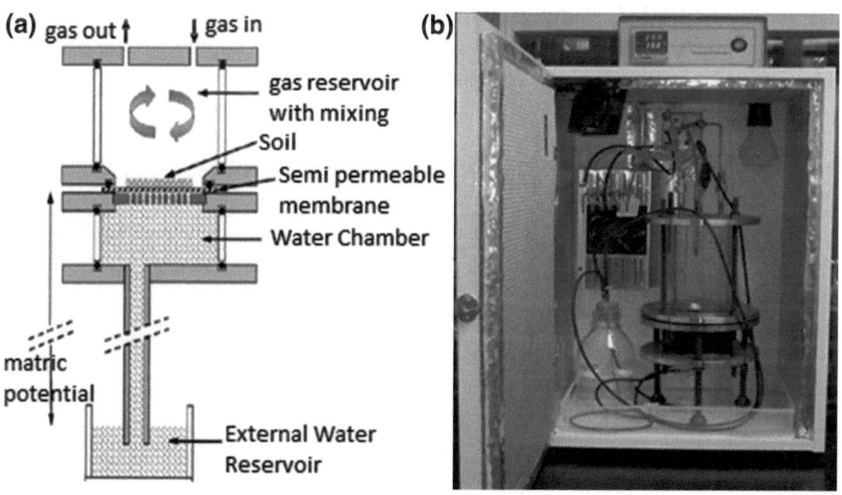

Fig. 1 **a** A cut-away of the differential biofiltration reactor with water content control, **b** experimental set-up

ratio of 11.2) is used as a bed medium for the current research. The reactor inlet and outlet toluene concentrations are monitored continuously by a gas chromatograph (GC) (SRI-8610C, SRI Instruments, CA, USA). In order to prevent condensation, the outlet line is heat traced at 40 °C and insulated until the GC sample entry port. In addition, a carbon dioxide analyser (GMP343, Vaisala Inc, CO, USA) is connected online to the GC sample purge port in order to measure the carbon dioxide production during the biodegradation of toluene (data not shown).

2.2 *Nutrient Addition*

New nutrient solutions (Table 1) are added by first draining the previous solution from both the upper and lower water chambers. Then the chambers are filled with phosphate buffered saline (PBS) at pH 7 to extract previous nutrient additions from the soil. This is repeated 3 times. Then the new nutrient solution is added to both chambers. All the nutrient solutions and the reactor at start-up are autoclaved at 121 °C for 45 min before being used in the experiment to minimise residual biological growth in the water reservoirs. Except for tap water and calcium chloride, all other nutrient solutions are prepared in the PBS solution at pH 7.0 in order to eliminate the pH effect on microbial degradation.

2.3 *Temperature Studies*

The temperature of the insulated box containing the differential reactor is controlled between 20 and 50 °C using a temperature controller (LTR-5, LAE Electronics, Italy) by a 60 W light bulb as a heat source; in addition, a cooling coil is added for obtaining temperatures below ambient.

Table 1 Different nutrients added to the differential biofilter to determine the impact on toluene elimination capacity

Nutrient	Concentration (g L^{-1})	Duration (days)
N source (a) $NaNO_3$	4.00 ± 0.005	5
P source (PB buffer) (a) KH_2PO_4 (b) $NaH_2PO_4 \cdot 2H_2O$	0.24 ± 0.001 1.44 ± 0.007	43 43
S, Mg, Fe sources (a) $MgSO_4 \cdot 7H_2O$ (b) $FeSO_4 \cdot 7H_2O$	0.2 ± 0.01 0.0008 ± 0.00001	7 7
Ca source (a) $CaCl_2 \cdot 2H_2O$	1.42 ± 0.005	56
Tap water	NA	15

2.4 Substrate Concentration Studies

The study of the toluene concentration effect on the elimination capacity (EC) ran for 4 months. Inlet concentrations were varied in a step-wise fashion. The inlet concentration was held constant until a steady outlet toluene concentration/EC was observed. Toluene inlet concentrations varied between 46.6 ± 0.5 ppm and 649.6 ± 4.2 ppm by varying the water bath temperature between 5 and 55 °C for this study. The experiment was repeated in three cycles with first two cycles at increasing order of concentrations and the last cycle at mixed order of concentrations.

3 Results and Discussion

3.1 Nutrient Effect

The differential biofilter reactor was operated initially for 7 days as an acclimation period for toluene degraders present in the biofilter medium (soil) with tap water in the water reservoirs. A steady EC was observed after the 7th day (Fig. 2), which was followed by the addition of the nutrient solutions in Table 1.

Following the steady state EC (31.9 ± 0.8 g m^{-3} h^{-1}) with tap water, at day 16, 0.01 M calcium chloride (Houba et al. 2000) replaced the tap water which slightly increased the EC to 35.9 ± 1.0 g m^{-3} h^{-1}. Since this marginal increase in EC was not considered significant, the toluene degraders present in the soil were not

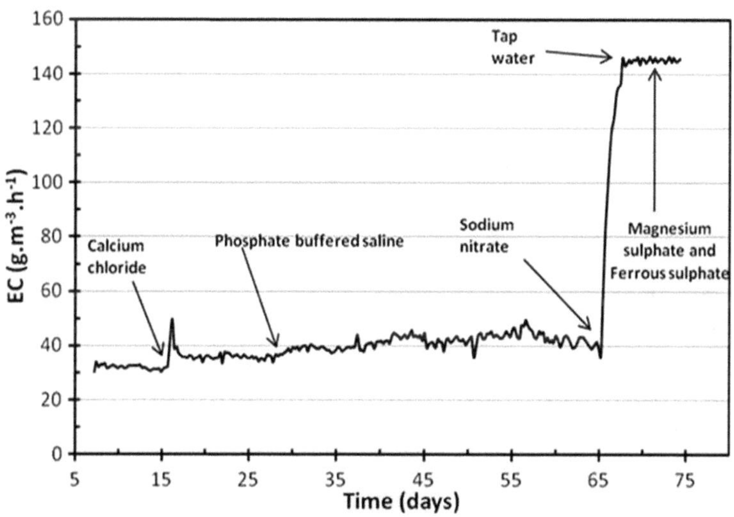

Fig. 2 Elimination capacity for nutrient addition studies at different time intervals

calcium-limited. After achieving steady state EC at day 29, phosphate buffered saline (PBS) was added by removing the calcium chloride from the reactor. Similar to calcium chloride, it slightly increased the EC after 10 days (day 39) to 41.5 ± 2.7 g m^{-3} h^{-1}. All subsequent experiments were performed with PBS buffer adjusted to 7.0. Moreover, PBS washes were performed whenever a new test solution was loaded and removed from the reactor. The pH of solution removed from the reactor over the rest of the subsequent experiments ranged from pH 6.5–6.8. After the day 66, PBS was replaced by 0.05 M sodium nitrate which increased the EC after 12 days to 145 ± 0.9 g m^{-3} h^{-1}, a 3.5-fold increase. This response indicated that nitrogen was the substrate limiting the growth of toluene degraders present in the biofilter bed. Nitrogen limitation has been commonly observed in other related biofiltration research work treating different volatile organic compounds (VOCs) (Acuña et al. 2002; Beuger and Gostomski 2009; Jorio et al. 2000; Morgenroth et al. 1996; Weckhuysen et al. 1993). After achieving a new steady state EC, the sodium nitrate solution was replaced with a PBS solution on day 69. No significant change in the EC was observed following this change. This response clearly indicated that the EC increase during the addition of sodium nitrate was due to biomass growth. On the 74th day, PBS was replaced with a magnesium sulphate/ferrous sulphate solution and no further change in EC was observed. This indicated that the toluene degraders were not Mg, Fe or sulphate limited.

3.2 *Temperature Effect*

The operating temperature of reactor was increased stepwise from 20 to 50 °C during the experiment and the EC increased gradually from 34 ± 2.0 g m^{-3} h^{-1} to a maximum of 49.8 ± 2.6 g m^{-3} h^{-1} at 45 °C (Fig. 3). However, the EC started to

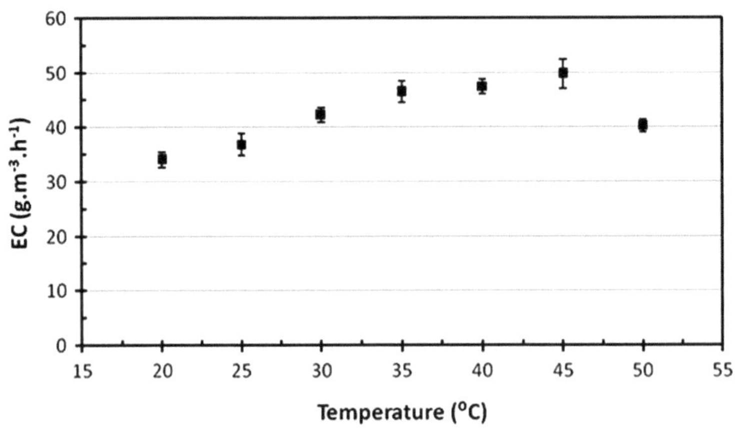

Fig. 3 Effect of temperature on elimination capacity

drop steeply above 45 °C. Similar studies reported that maximum specific toluene degradation rate was observed at 30 °C in a biofilter (Lee et al. 2002; Vergara-Fernández et al. 2007; Yoon and Park 2002). Hence from the current study it can be concluded that a highest intensity of the metabolic microbial activity in soil was seen at 45 °C which was 18 % lower than the optimal temperature reported earlier by Beuger and Gostomski (2009) in compost.

3.3 Substrate Concentration Effect

The effect of residual (outlet) toluene concentration on EC was studied by manipulating the inlet toluene concentrations to change the load. Before these experiments were started, the reactor was operated for 120 days, and any excess nutrients to stimulate growth were assumed to be exhausted. Therefore it was assumed that maintenance energy was driving the toluene degradation with no net change in active biomass. Three cycles of repeat experiments were conducted and in each experiment the EC increased with residual substrate concentration before dropping at higher toluene concentrations (Fig. 4).

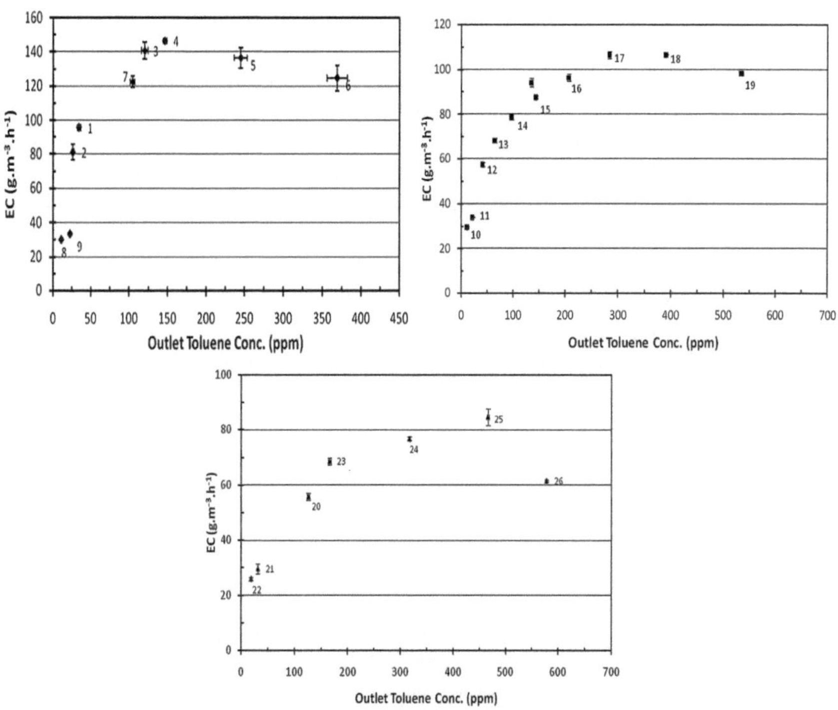

Fig. 4 The relationship between the outlet (residual) toluene concentration and the EC for cycle 1–3

The initial increase in EC with increasing toluene concentrations seen in all three cycles is a typical response of a biofilm combining diffusion-limited degradation at lower concentrations followed by full biofilm penetration at higher concentrations as described by Ottengraf and van den Oever (1983). The subsequent drop in EC at higher toluene concentrations seen in all three cycles indicates substrate inhibition (Zilli et al. 2000). The other outstanding anomaly is that the biofilter displayed a steady EC of approximately 38 g m^{-3} h^{-1} for over 100 days at 160 ppm prior to the start of cycle 1. However, after a relatively brief time at lower residual concentrations (#1, #2), an increase back to the range of 160 ppm gave a much higher EC ($\sim$145 g m^{-3} h^{-1}, #4). At present there is no acceptable explanation for this and is probably related to inhibition and how different strains in the microbial community responded overtime to different toluene concentrations and is the subject of further.

4 Conclusion

The differential biofilter reactor used in these studies showed a high degree of flexibility in manipulating environmental parameters, such as substrate concentration, dissolved nutrients and operating temperature while controlling water content. From the substrate concentration studies, it was demonstrated that below critical substrate concentration, mass transfer limitation influenced the EC and above that, substrate inhibition was the dominant influence. Furthermore, by replicate substrate concentration studies on the same biofilter media (soil) demonstrated a loss of active toluene degraders but less substrate inhibition. This drop in EC was attributed to exposure to higher toluene concentrations during earlier cycles. Studies carried out using different nutrients clearly proved that the toluene degraders present in the soil were nitrogen limited. This was evident from the 3.5-fold increase in the EC with the addition of nitrogen ($NaNO_3$) but the other nutrients did not show a significant increase in EC. Temperature studies showed that the EC of the soil in the differential biofilter reactor increased with increasing temperature, from 34 ± 1.4 g m^{-3} h^{-1} to 49.8 ± 2.6 g m^{-3} h^{-1} for temperatures of 20–45 °C, respectively and dropping above 45 °C.

References

Abdel-El-Haleem, D.: Acinetobacter: environmental and biotechnological applications. Afr. J. Biotechnol. **2**(4), 71–74 (2004)

Acuña, M.E., Villanueva, C., Cardenas, B., Christen, P., Revah, S.: The effect of nutrient concentration on biofilm formation on peat and gas phase toluene biodegradation under biofiltration conditions. Process Biochem. **38**(1), 7–13 (2002)

Alahari, A., Apte, S.K.: A novel potassium deficiency-induced stimulon in *Anabaena torulosa*. J. Biosci. **29**(2), 153–161 (2004)

Andrews, J.F.: A mathematical model for the continuous culture of microorganisms utilizing inhibitory substrates. Biotechnol. Bioeng. **10**(6), 707–723 (1968)

Beuger, A.L., Gostomski, P.A.: The Impact of Environmental Parameters on Toluene Degradation Using a Laboratroy-Scale Reactor with Internal Recycle. Engineering Our Future: Are We up to the Challenge, pp. 303–307, (2009)

Detchanamurthy, S., Gostomski, P.A.: Development of a modified differential biofiltration reactor with online sample and carbon dioxide monitoring system. Asia-Pac. J. Chem. Eng. **8**(3), 414–424 (2012)

Doran, P.M.: Bioprocess Engineering Principles (vol. 1). Academic Press, Cambridge (1995)

Du Plessis, C.A., Kinney, K.A., Schroeder, E.D., Chang, D.P.Y., Scow, K.M.: Denitrification and nitric oxide reduction in an aerobic toluene-treating biofilter. Biotechnol. Bioeng. **58**(4), 408–415 (1998)

Houba, V., Temminghoff, E., Gaikhorst, G., Van Vark, W.: Soil analysis procedures using 0.01 M calcium chloride as extraction reagent. Commun. Soil Sci. Plant Anal. **31**(9–10), 1299–1396 (2000)

Jorio, H., Bibeau, L., Heitz, M.: Biofiltration of air contaminated by styrene: effect of nitrogen supply, gas flow rate, and inlet concentration. Environ. Sci. Technol. **34**(9), 1764–1771 (2000)

Kennes, C., Veiga, M.C.: Bioreactors for Waste Gas Treatment (vol. 4). Kluwer Academic Publishers, Berlin (2001)

Kiared, K., Fundenberger, B., Brzezinski, R., Viel, G., Heitz, M.: Biofiltration of air polluted with toluene under steady-state conditions: experimental observations. Ind. Eng. Chem. Res. **36**(11), 4719–4725 (1997)

Lee, E.Y., Jun, Y.S., Cho, K.S., Ryu, H.W.: Degradation characteristics of toluene, benzene, ethylbenzene, and xylene by *Stenotrophomonas maltophilia* T3-c. J. Air Waste Manag. Assoc. **52**(4), 400–406 (2002)

Leson, G., Winer, A.M.: Biofiltration—an innovative air-pollution control technology for VOC emissions. J. Air Waste Manag. Assoc. **41**(8), 1045–1054 (1991)

Morgenroth, E., Schroeder, E.D., Chang, D.P.Y., Scow, K.M.: Nutrient limitation in a compost biofilter degrading hexane. J. Air Waste Manag. Assoc. **46**(4), 300–308 (1996)

Nikiema, J., Brzezinski, R., Heitz, M.: Influence of phosphorus, potassium, and copper on methane biofiltration performance (A paper submitted to the journal of environmental engineering and science). Can. J. Civ. Eng. **37**(2), 335–345 (2010)

Ottengraf, S.P.P., Vandenoever, A.H.C.: Kinetics of organic-compound removal from waste gases with a biological filter. Biotechnol. Bioeng. **25**(12), 3089–3102 (1983)

Sorial, G.A., Smith, F.L., Suidan, M.T., Pandit, A., Biswas, P., Brenner, R.C.: Evaluation of trickle-bed air biofilter performance for styrene removal. Water Res. **32**(5), 1593–1603 (1998)

Vergara-Fernández, A., Lara Molina, L., Pulido, N.A., Aroca, G.: Effects of gas flow rate, inlet concentration and temperature on the biofiltration of toluene vapors. J. Environ. Manag. **84**(2), 115–122 (2007)

Weckhuysen, B., Vriens, L., Verachtert, H.: The effect of nutrient supplementation on the biofiltration removal of butanal in contaminated air. Appl. Microbiol. Biotechnol. **39**(3), 395–399 (1993)

Yoon, I.K., Park, C.H.: Effects of gas flow rate, inlet concentration and temperature on biofiltration of volatile organic compounds in a peat-packed biofilter. J. Biosci. Bioeng. **93**(2), 165–169 (2002)

Zilli, M., Del Borghi, A., Converti, A.: Toluene vapour removal in a laboratory-scale biofilter. Appl. Microbiol. Biotechnol. **54**(2), 248–254 (2000)

Importance of Biomass-Specific Pretreatment Methods for Effective and Sustainable Utilization of Renewable Resources

Yadhu N. Guragain and Praveen V. Vadlani

1 Introduction

Transportation fuels and climate change are topics of keen discussion in political and scientific communities throughout the world. A gradual decrease in economically and sustainably recoverable petroleum reserves requires a search for alternative to fossil fuels. In addition, increasing instability in most of the oil-producing countries has further worsened the fuels crisis for the non-oil-producing countries (Mousdale 2008). Bio-based fuel and chemical production has rapidly increased in last one and half decades; however, current biorefineries predominantly use food-based materials as feedstocks (Guragain et al. 2016b), which is not sustainable due to a global food security issue. The abundantly available lignocellulosic residues are sustainable feedstocks for bio-based fuels and chemicals production without affecting global food supply (Guragain 2015a).

Bioprocessing of lignocellulosic biomass can be broadly divided into three unit operations: pretreatment, hydrolysis and fermentation, as shown in Fig. 1. Each step of this bioconversion process is associated with a number of challenges (Mousdale 2008); among them pretreatment is the central unit operation that significantly affects the efficiency of all subsequent steps of bioprocessing, and is the most expensive single unit operation in the context of existing biomass conversion technologies (Guragain 2015a). In addition, valorization of lignin stream is critical for the commercial viability of biorefinery industries (Ragauskas et al. 2014);

Y.N. Guragain (✉) · P.V. Vadlani
Bioprocessing and Renewable Energy Laboratory Grain Science and Industry Department, Kansas State University, Manhattan, KS, USA
e-mail: guragain@ksu.edu

P.V. Vadlani
Department of Chemical Engineering, Kansas State University, Manhattan, KS, USA

Prasanna B.D. et al. (eds.), *Biotechnology and Biochemical Engineering*,
DOI 10.1007/978-981-10-1920-3_23

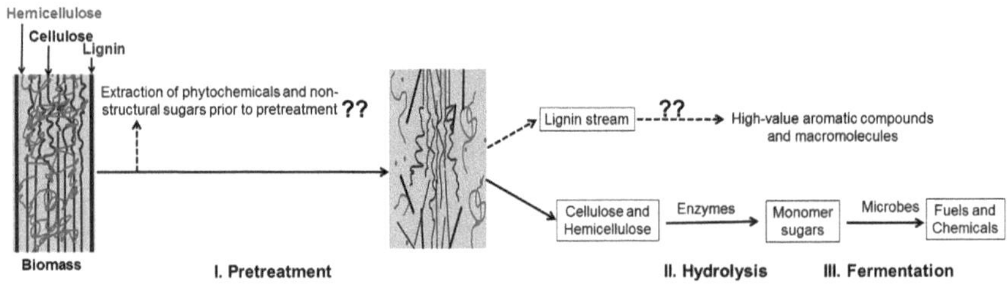

Fig. 1 Bioprocessing for lignocellulosic biomass to produce fuels and chemical

biomass pretreatment significantly effects the quality of extracted lignin for high value application.

Pretreatment methods are broadly divided into physical, chemical, physico-chemical, and biological pretreatment methods. Each pretreatment method is associated with some advantages as well as disadvantages (Guragain 2015a). In addition, variations in biomass composition among different biomass feedstocks significantly affect the effectiveness of each pretreatment method. Moreover, structure of biopolymers, especially lignin, differs from biomass to biomass (Guragain et al. 2015b), which further affect the pretreatment efficiency. Biomass lignin is made up of three types of monomer units such as syringyl (S), guaiacyl (G) and *p*-hydroxyphenyl (H). In addition, grass lignin also contains a significant proportion of *p*-coumarate and ferulate monomers (Guragain et al. 2015b) (Fig. 2). The proportion of different lignin monomers, and type of linkages between these monomers in biomass lignin affect delignification efficiency during pretreatment process. The pretreatment efficiency is better for biomass containing higher S/G ratio, higher proportion of *p*-coumarate and ferulate monomers, and higher amount of ester and ether inter-unit linkages, whereas the efficiency decreases with increase in carbon-carbon linkages (Studer et al. 2011).

In this study, a wide range of biomass feedstocks, including grass, softwood and hardwood, were evaluated and compared for the most common biomass pretreatment methods, such as dilute acid and alkali pretreatment. These biomass feedstocks were also evaluated for the effectiveness of organic solvent pretreatment using alkali catalyst. Alkaline organosolv pretreatment is considered as the promising pretreatment method to extract good quality biomass lignin to produce high-value products from the extracted lignin (Guragain et al. 2016a). Effectiveness of these pretreatment methods were evaluated based on the sugar released during enzymatic hydrolysis of pretreated biomass. In addition, three sorghum cultivars (Early Hegari, Atlas and Kansas Collier) and brown midrib (*bmr*)-12 mutant of each cultivar were evaluated for sugar release during enzymatic hydrolysis of alkali pretreated biomass. The objective of *bmr* mutation is to perturb lignin biosynthesis pathways leading to change in total lignin content and/or proportion of different type of lignin monomers so that a new crop line with better feedstock quality for biofuels and biochemical production is developed (Guragain 2015a).

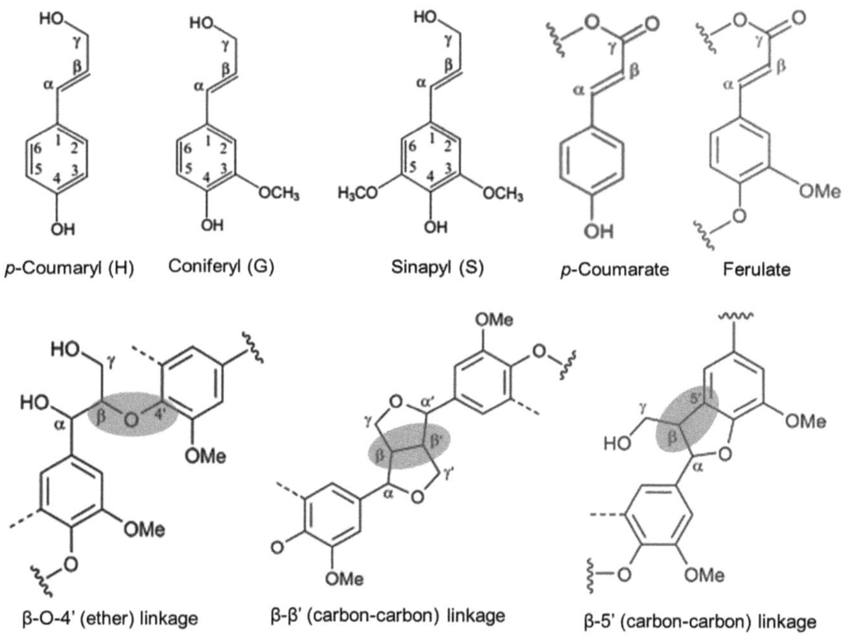

Fig. 2 Lignin monomers and inter-lignin linkages in biomass lignin structure (Guragain et al. 2015b)

2 Materials and Methods

A schematic representation of the methodology used in this study is given in Fig. 3. Biomass composition was first determined using standard protocols (Guragain 2015a). Dilute acid and alkali pretreatment methods were evaluated and compared for a wide variety of biomass feedstocks such as grass (corn stover, switchgrass, and sorghum stalks), softwood (Douglas fir) and hardwood (poplar). In addition, alkaline organic solvent pretreatment method was evaluated using two types of organic solvents (high- and low-boiling point solvents) for corn stover, Douglas fir and poplar. Glycerol and 2,3-butanediol were used as high-boiling point solvents, and isopropanol, ethanol, butanol and water were used as low-boiling point solvents. Pretreatment conditions were optimized for processing temperature and catalyst (sodium hydroxide) concentration; temperature range was less than the boiling point of solvent for atmospheric pressure processing. The pretreated slurry was filtered, washed and residual solid was enzymatically hydrolysed using the methods used by Guragain et al. (2013). Monomers sugars were measured using high performance liquid chromatography (Shimadzu Corporation, Japan), and data from triplicate experiments were statistically analysed for Least Significant Difference (LSD) at 95 % confidence level ($P < 0.05$) using JMP software (SAS Institute Inc., Cary, North Carolina, United States).

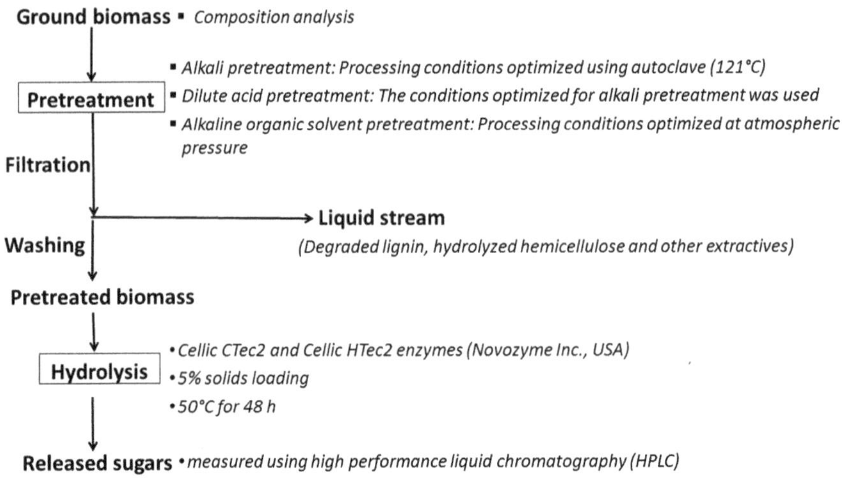

Fig. 3 Schematic representation of biomass processing. Sulfuric acid and sodium hydroxide (NaOH) were used in acid and alkali pretreatment process, respectively. Organosolv pretreatment using alkali (NaOH) catalyst was optimized separately for high- and low-boiling point solvents using corn stover as feedstock, and the optimized conditions were used for all biomass samples

In a separate experiment, three sorghum cultivars (Early Hegari, Atlas and Kansas Collier) and *bmr*12 mutant of each cultivar were evaluated and compared for non-structural content and sugar released from structural carbohydrates (cellulose and hemicellulose). The non-structural sugar content was determined by extracting ground biomass in water using Soxhlet extraction, and the sugar released from structural carbohydrates was determined by alkali pretreatment of ground biomass followed by enzymatic hydrolysis. Total fermentable sugar was then calculated by adding these two types of sugars: non-structural sugars and released sugars from structural carbohydrates.

3 Results and Discussion

Figure 4 shows that biomass composition significantly varied among different biomass samples; the composition variation was not only among softwood (Douglas fir), hardwood (poplar) and grass (corn stover, switchgrass and sorghums), but also among different grass samples. The acid pretreatment was significantly less effective than alkali pretreatment at the same processing conditions, including processing time, temperature and acid/alkali concentration, to release sugar during enzymatic hydrolysis for all biomass feedstocks. Alkali pretreatment delignifies biomass, making the carbohydrate polymers accessible for enzyme hydrolysis; whereas, acid pretreatment expose cellulose for cellulase enzyme accessibility by hydrolysing hemicellulose (Leu and Zhu 2013). The results

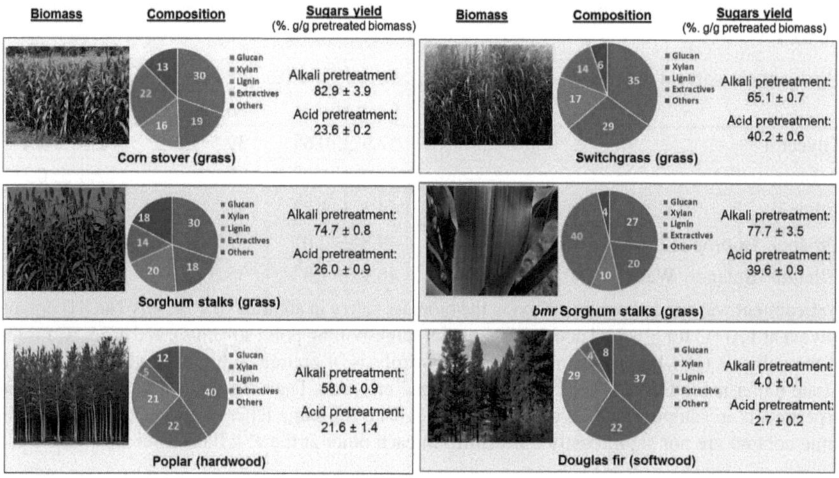

Fig. 4 Composition of different biomass feedstocks, and sugar released during enzymatic hydrolysis after alkali (NaOH) and acid (H_2SO_4) pretreatment at same processing conditions: 121 °C for 30 min using 1 and 2 % acid (v/v)/alkali (w/v) for grass and woody biomass, respectively. *bmr* = brown midrib sorghum mutant. Hydrolysis of pretreated biomass was carried out in citrate buffer using Cellic CTec2 and Cellic HTec2 enzymes. Composition data are average values of triplicate experiments, and the sugar yield data are average values of triplicate experiments ± sample standard deviation

indicated that harsher conditions are required for hemicellulose hydrolysis by acid process compared to lignin degradation by alkali. The results also showed that acid pretreatment effectiveness increase with increase in hemicellulose content in extractives-free basis, indicating that structural variation in hemicellulose is not significant among different biomass samples. On the other hand, decrease in total lignin in biomass was not necessarily beneficial to increase alkali pretreatment effectiveness with the same proportion. For example, sugars released from alkali pretreated corn stover was 27 % more than switchgrass while both biomass had statistically equal lignin content; similar results were observed between sorghum stalks and poplar. Moreover, a decrease in lignin content by 50 % in *bmr* sorghum compared to wild type sorghum resulted in an increase in sugar yield only by 3 %. These results are consistent with the facts that S/G ratio in corn stover lignin is more than that of switchgrass lignin (Guragain et al. 2015b), thereby making corn stover more susceptible for alkali pretreatment. Poplar lignin contains higher S/G ratio compared to grass lignin; however, absence of *p*-coumarate and ferulate lignin monomers in poplar lignin polymer (Guragain et al. 2015b) indicate that the poplar lignin contains significantly lower proportion of alkali labile ether linkages, and higher proportion of strong carbon-carbon linkages compared to grass lignin. These differences in lignin structure could be the reason for a less alkali pretreatment effectiveness in poplar than a grass (sorghum stalks) with same amount of total lignin content. High total lignin (29 %) and low extractives (4 %) coupled with

Table 1 Sugar released during enzymatic hydrolysis of alkaline organic solvent pretreated biomass

Pretreatment solvent system	Sugar yield (%, g/g raw biomass)		
	Corn stover	Poplar	Douglas fir
Glycerol	37.9 ± 0.6[c]	39.5 ± 0.2[a]	2.3 ± 0.1[a]
2,3-butanediol	43.9 ± 6.0[bc]	38.9 ± 0.2[b]	2.4 ± 0.2[a]
Ethanol	44.8 ± 3.8[b]	32.3 ± 2.0[d]	1.9 ± 0.5[a]
Ethanol: Isopropanol = 1:1 (by volume)	49.8 ± 2.0[a]	31.2 ± 1.2[d]	2.0 ± 0.3[a]
Ethanol: Butanol: Water = 1:1:1 (by volume)	46.9 ± 1.8[b]	35.0 ± 0.6[c]	2.1 ± 0.3[a]

Pretreatment was carried out by refluxing the biomass slurry in alkaline (0.4 %, w/v NaOH) organic solvent at 170 °C for glycerol and 2,3-butanediol (high-boiling point solvents), and at 75 °C for rest of the solvents (low-boiling points solvents). Hydrolysis of pretreated biomass was carried out in citrate buffer using Cellic CTec2 and Cellic HTec2 enzymes. Data are average values of triplicate experiments ± sample standard deviation. Values with the same letters in superscripts within the same column are not significantly different from each other at the $P < 0.05$ level

high proportion of C–C linkages in softwood (Douglas fir) lignin was the reason for ineffective pretreatment in both acid and alkali processes for this biomass sample.

Table 1 shows that mixture of low-boiling point alcohols were better solvent system for corn stover pretreatment than high-boiling point solvents, whereas opposite was true for poplar; glycerol was the promising solvent for poplar pretreatment. None of these solvents worked well for Douglas fir pretreatment. These data indicated that an appropriate alkaline organic solvent for pretreatment (delignification) of any biomass feedstock depends on biomass types due to a distinct variation in composition (Fig. 4) and structure of biomass lignin (Guragain et al. 2015b). Therefore, each biomass feedstock must be separately optimized for each organic solvent to develop an effective organosolv pretreatment process.

We also evaluated and compared three sorghum cultivars (Early Hegari, Atlas and Kansas Collier) and *bmr* 12 mutant of each cultivar. Table 2 shows that *bmr* 12 mutants of Atlas and Kansas Collier had 36.2 and 30.1 % more total fermentable sugars, respectively, than their parent cultivars, whereas *bmr* 12 of Early Hegari had 12.1 % less total fermentable sugar than its parent line. Total sugar released during enzymatic hydrolysis was significantly higher in all *bmr* mutants than their parent cultivars, indicating that *bmr* mutation led to improved hydrolysis efficiency of sorghum biomass. Sugar yield per unit weight of sorghum biomass was further increased in *bmr* mutants compared to their parents because of higher amount of non-structural sugars (glucose, fructose and sucrose) in Atlas and Kansas Collier cultivars. However, the benefits from improved hydrolysis efficiency in *bmr* 12 mutant of Early Hegari was surpassed by decreased non-structural sugar content in this mutant compared to its parent line. These results showed that *bmr* 12 mutation leads to better feedstocks for biofuels and biochemical production in some sorghum cultivars, while the same mutation has adverse effects in other cultivar. Therefore, *bmr* mutation of each sorghum cultivar must be separately evaluated to develop a promising bioenergy crops.

Table 2 Sugar released during enzymatic hydrolysis of alkaline organic solvent pretreated biomass

Sorghum genotype	Non-structural sugar (g/100 g biomass)	Sugar released in hydrolysis (g/100 g biomass)	Total fermentable sugar (g/100 g biomass)
Early Hegari (EH)	15.40 ± 0.12^{d}	34.34 ± 0.80^{b}	49.74 ± 0.78^{c}
EH *bmr* 12 mutant	2.70 ± 0.00^{f}	40.84 ± 1.18^{a}	43.54 ± 1.18^{e}
Atlas (AT)	11.10 ± 0.10^{e}	30.51 ± 0.95^{c}	41.61 ± 0.93^{f}
AT *bmr* 12 mutant	21.40 ± 0.14^{b}	35.27 ± 1.19^{b}	56.67 ± 1.26^{b}
Kansas Collier (KC)	16.69 ± 0.03^{c}	29.32 ± 1.46^{c}	46.01 ± 1.43^{d}
KC *bmr* 12 mutant	26.34 ± 0.46^{a}	33.52 ± 0.42^{b}	59.86 ± 0.68^{a}

bmr = brown midrib sorghum mutant. Non-structural sugars was total sugars extracted from ground biomass in water using Soxhlet extraction. Sugar released in hydrolysis was total sugars generated during enzymatic hydrolysis of pretreated biomass. Pretreatment was carried out at 121 °C for 30 min using 1.25 % (w/v) sodium hydroxide solution followed by filtration and enzymatic hydrolysis of residual solid was carried out in citrate buffer using Cellic CTec2 and Cellic HTec2 enzymes. Data are average values of triplicate experiments ± sample standard deviation. Values with the same letters in superscripts within the same column are not significantly different from each other at the $P < 0.05$ level

As a consequence, above results show that effectiveness of different pretreatment methods differ from biomass to biomass. A number of factor other than gross biomass composition significantly affect the saccharification efficiency of pretreated biomass, including lignin structure and crystallinity of cellulose (Guragain et al. 2014). Therefore, a fundamental understanding of gross biomass composition, structure of each biopolymer, and the interactions among different biopolymers within a biomass is critical to develop an appropriate pretreatment method specific to each biomass feedstock. In addition, plant breeding, such as *bmr* mutation, to develop better feedstocks for biofuels and biochemical production should be separately evaluated for each plant cultivar because of a distinct variation on the effect of same mutation on different cultivars of a plant crop.

4 Summary and Conclusions

This study showed that a single pretreatment method cannot be an all-encompassing process for different types of biomass feedstocks due to distinct variations in the effectiveness among various pretreatment methods. Structure of biopolymers and their interactions within biomass are critical, in addition to gross biomass composition, in the development of an appropriate pretreatment method specific to each biomass feedstock. In addition, biomass pretreatment process must be tailored with the optimum use of each component of biomass, including lignin. For example, alkali pretreatment process solubilizes biomass lignin, opening up an opportunity to produce high-value phenolic compounds. Whereas acid pretreatment process solubilizes hemicellulose so that biomass lignin can be obtained after hydrolysis of cellulose from pretreated biomass, which can be used for composite materials production. Moreover, separate cellulose and hemicellulose hydrolyzates can be obtained in acid pretreatment process because hemicellulose and cellulose are hydrolyzed during pretreatment and enzymatic hydrolysis steps, respectively, in this process. A separate biomass-derived glucose and xylose sugar is beneficial for efficient fermentation to fuel and chemicals. Nevertheless, harsher pretreatment conditions are required in acid pretreatment compared to alkali pretreatment, leading to a need for additional detoxification process to generate clean sugars. Less pretreatment-induced inhibitory compounds are produced in alkali pretreatment compared to acid pretreatment due to the milder processing conditions; however, both cellulose and hemicellulose are hydrolyzed during enzymatic hydrolysis step leading to a mixture of glucose and xylose sugars in hydrolyzate, which is less efficient medium for fermentation compared to single sugar medium (Guragain 2015a). Furthermore, some biomass feedstocks, such as sorghum stalks, contain as high as 20 % (w/w) of non-structural sugars, including sucrose, glucose, and fructose, which must be extracted prior to biomass pretreatment to avoid their degradation during pretreatment process. Similarly, separate innovative strategies are required for the valorization of various phytochemicals such as terpenes, fats, waxes, phenolics, and alkaloids. However, all biomass feedstocks do not contain these non-structural sugars and phytochemicals in economically feasible level; therefore, a targeted biorefining strategy is required for each type of biomass feedstock.

Acknowledgments This work was funded by the Development Initiative Competitive Grants Program (BRDI; grant number: 2012-10008-20263), and the Small Business Innovation Research (SBIR), Department of Navy; Grant Number: N68335-13-C-0174. Author PVV thanks the Lortscher Endowment for their support. The authors are also grateful to Novozymes Inc. for enzyme samples.

References

Guragain, Y.N.: Sustainable bioprocessing of various biomass feedstocks: 2,3-butanediol production using novel pretreatment and fermentation. Doctoral Dissertation, Kansas State University, USA (2015a)

Guragain, Y.N., Wilson, J., Staggenborg, S., McKinney, L., Wang, D., Vadlani, P.V.: Evaluation of pelleting as a pre-processing step for effective biomass deconstruction and fermentation. Biochem. Eng. J. **77**, 198–207 (2013)

Guragain, Y.N., Ganesh, K., Bansal, S., Sathish, R.S., Rao, N., Vadlani, P.V.: Low-lignin mutant biomass resources: effect of compositional changes on ethanol yield. Ind. Crop Prod. **61**, 1–8 (2014)

Guragain, Y.N., Bastola, K.P., Madl, R.L., Vadlani, P.V.: Novel biomass pretreatment using alkaline organic solvents: a green approach for biomass fractionation and 2,3-butanediol production. BioEnerg. Res. **9**, 643–655 (2016a). doi:10.1007/s12155-015-9706-y

Guragain, Y.N., Herrera, A.I., Vadlani, P.V., Prakash, O.: Lignins of bioenergy crops: a review. Nat. Prod. Commun. **10**, 201–208 (2015b)

Guragain, Y.N., Probst, K.V., Vadlani, P.V.: Fuel alcohol production. In: Corke, H., Faubion, J., Seethraman, K., Wrigley, C. (eds.) Encyclopedia of Food Grain, pp. 235–244. Elsevier, Oxford (2016b)

Leu, S., Zhu, J.: Substrate-related factors affecting enzymatic saccharification of lignocelluloses: our recent understanding. Bioenerg. Res. **6**, 405–415 (2013)

Mousdale, D.M.: Biofuels: Biotechnology, Chemistry, and Sustainable Development. CRC Press, Boca Rato (2008)

Ragauskas, A.J., Beckham, G.T., Biddy, M.J., Chandra, R., Chen, F., Davis, M.F., et al.: Lignin valorization: improving lignin processing in the biorefinery. Science **344**(6185), 1246843 (2014)

Studer, M.H., DeMartini, J.D., Davis, M.F., Sykes, R.W., Davison, B., Keller, M., et al.: Lignin content in natural *populus* variants affects sugar release. Proc. Natl. Acad. Sci. U.S.A. **108**, 6300–6305 (2011)

Medium Optimisation for Maximum Growth/Biomass Production of *Arthrobacter sulfureus* for Biodesulphurisation

E. Asha Rani, M.B. Saidutta and B.D. Prasanna

1 Introduction

Diesel oil contains large amount of organosulphur compounds, combustion of which leads to Sulphur oxide emissions causing harmful effects on environment. Hydrodesulphurisation (HDS), a conventional technique employed by refineries operates at high temperature and pressure, but this process is being ineffective in removal of organic sulphur compounds like dibenzothiophenes (DBT), therefore it is not applicable for deep-desulphurisation. In this case, Biodesulpharisation (BDS) as a complement to deep HDS could be an interesting option for removal of remaining sulphur compounds. BDS is a promising technology which works at mild operating conditions and is considered "green" (Luo et al. 2003).

Arhtrobacter sulfureus is capable of removing sulphur from DBT in a selective way without carbon skeleton rupture. Since the bacteria transform the DBT molecule into 2-hydroxybiphenyl (HBP) and sulphate by means of a non-destructive pathway, usually called 4S pathway by which fuel value is preserved (Labana et al. 2005). Other microorganisms such as, *Rhodococcus* sp. Strain T09 (Matsui et al. 2001), *Corynebacterium* sp. Strain SY1 (Omori et al. 1992), *Pseudomonas delafieldii* R-8 (Shan et al. 2005) have been reported which helps in removal of sulphur from this kind of compounds. For BDS process, initially the organism is grown by employing nutrient source and after a desulphurising biomass is obtained; these cells in the form

E. Asha Rani (✉) · M.B. Saidutta · B.D. Prasanna
Department of Chemical Engineering, National Institute of Technology Karnataka, Surathkal, Srinivasnagar 575025, Karnataka, India
e-mail: asha.devnekar@gmail.com

M.B. Saidutta
e-mail: saidutta.mb@gmail.com

B.D. Prasanna
e-mail: prsnbhat@gmail.com

Prasanna B.D. et al. (eds.), *Biotechnology and Biochemical Engineering*,
DOI 10.1007/978-981-10-1920-3_24

of resting cellsare employed to remove sulphur from diesel and model oil in a biphasic system.

For efficient production of desulphurising cells, it is highly essential to optimise the media composition which further helps in economic design of large scale fermentation operation system, making it cost effective which is one of the drawbacks of BDS process. Central Composite Design (CCD) and Box-Behnken Design (BBD) of RSM are well known fractional factorial designs used for optimization of variables with a limited number of experiments (Hasan et al. 2010; Wu et al. 2010).

The aim of this work is to optimise medium components for production of *Arthrobacter sulfureus* with high biomass yield using CCD and to understand the influence of operational conditions (concentration of cells, water-oil ratio and substrate concentration) for treatment of diesel and model oil in a biphasic system using resting cell suspension.

2 Materials and Methods

High-speed Diesel oil was obtained from Mangalore Refinery and Petrochemicals Limited, India.NB (Nutrient Broth), a complex medium is used for preparing an inoculum of *A. Sulfureus* and a standard medium (BSM) is used for the optimization of growth media which has following composition:

$NaH_2PO_4 \cdot H_2O$ 4 g/L, $K_2HPO_4 \cdot 3H_2O$ 4 g/L, $CaCl_2 \cdot 2H_2O$ 0.001 g/L $MgCl_2 \cdot 6H_2O$ 0.0245 g/L, $FeCl_3 \cdot 6H_2O$ 0.001 g/L in addition with 2 % carbon source (glucose and sodium succinate), nitrogen source (ammonium chloride and ammonium nitrate), sulphur source (DBT, DMSO and $MgSO_4$) were added.
Resting cell suspension is prepared for biphasic reaction by harvesting cells during mid-log phase and suspending the harvested cells in phosphate buffer with 2 % glucose which acts as an aqueous phase. The procedure for harvesting cell is by centrifuging at 10,000 rpm for 10 min at 4 °C and washed twice with 0.1 M phosphate buffer of pH 7.
After 5 days of incubation of biphasic system, the culture was centrifuged at 10,000 g for 10 min to separate the oil and aqueous phases. The total sulphur content in the oil phase was measured by X-ray fluorescence method and in accordance with ASTM standard D-4294.

3 Results and Discussion

3.1 Optimisation of Biomass Growth

One-factor-at-a-time method was adopted to analyse the effect of medium components (carbon, nitrogen and sulphur components) on biomass growth in our earlier experiments. Different nutrient sources likecarbon source (Glucose and sodium succinate), nitrogen source (ammonium chloride and ammonium nitrate) and sulphur source (DBT, DMSO and magnesium sulphate) have been checked among which glucose, ammonium chloride and DMSO were selected as the best carbon, nitrogen and sulphur source giving higher growth rate. Glucose is the most commonly employed carbon source in the growth of various BDS strains like in *Gordona strain* CYKS1 (Rhee et al. 1998) and *Rhodococcus erythropolis* KA2-5-1 (Yan et al. 2000). Similarly, most commonly employed nitrogen source is found to be ammonium chloride (Shan et al. 2005; Yu et al. 2006). The growth rate of *Arthrobacter sulfureus* on DMSO was higher compared to DBT being used as the sole sulfur source and in further experiments it was observed that the resting cells of *Arthrobacter sulfureus* grown on DMSO as the sulphur source were more effective in desulfurizing DBT. A similar observation was reported by Mohebali et al. (2008).

RSM was used for better understanding of interaction between medium components and further optimisation for higher biomass production of *Arthrobacter sulfureus*. In order to define the optimal response region of biomass production, experimental values of biomass yield were subjected to multiple linear regression analysis using MINITAB 16.

Preliminary trials were carried out to set the concentration range for these components and significance of regression the lowest and highest level of variables considered for RSM studies is given in Table 1. The effect of each compound on biomass production was taken as a response implemented in 20 experiments wherein 2^3 factorial design was augmented by 6 axial points ($\alpha = 1.682$) and three replicates at the center point.

The effect of glucose, ammonium chloride and DMSO on biomass production (g/L) was described in the form Eq. (1) and a second order polynomial model in coded unit of the variables.

$$\begin{aligned}\text{Biomass yield} =& (6.82507) + (0.132329)A + (1.55278)B + (0.00536844)C - (0.00141858)A^2 \\ & - (0.159358)B^2 - (2.91708E-06)C^2 - (0.0105556)AB \\ & + (1.08333E-05)AC - (3.141667E-04)BC\end{aligned} \tag{1}$$

where variable A is Glucose concentration (g/L), variable B is ammonium chloride concentration (g/L) and variable C is DMSO concentration (μM).

Table 2 shows the analysis of variance (ANOVA), coefficient determination (R^2), adjusted determination for the experiment. The coefficient determination (R^2)

Table 1 Experimental range and design matrix

Component	−α	−1	0	+1	+α
Glucose (g/l)	0.505	10	25	40	49.495
Ammonium chloride (g/l)	0.0505	1	2.5	4	4.9495
DMSO (μM)	73.4	200	400	600	726.6

Table 2 ANOVA for response surface quadratic model

Source	DF	Seq SS	Adj SS	Adj MS	F	P
Regression	9	15.0930	15.0930	1.67700	48.17	0.000
Linear	3	11.6392	11.6392	3.87973	111.43	0.000
Square	3	2.9100	2.9100	0.97001	27.86	0.000
Interaction	3	0.5437	0.5437	0.18125	5.21	0.023
Lack-of-fit	5	0.2294	0.2294	0.04588	2.19	0.234
Pure error	4	0.0839	0.0839	0.02098	0	0
Total	19	15.4090	0	0	0	0

was shown as 0.9797, indicating 97.97 % of the variability in the response, while adjusted determination coefficient value was Adj. R2 = 0.95 for which the model effectively explains and advocates for significance $R^2 = 97.97$ %, R^2(pred) = 84.95 %, R^2(adj) = 95.71 %.

3.2 *Analysis of Response Surface*

The correlation between variables is expressed as surface plots and counter plots. Figure 1 describes the response surface plots which show the interaction effect of glucose, ammonium chloride and DMSO concentration on the biomass growth. In these plots, one factor remains constant at the optimum level whereas the other two factors vary within their experimental range.

The model was then subjected to optimization using "Response optimizer" of MINITAB 16 and the optimized biomass yield (y) was found to be 13.31 g/L. The

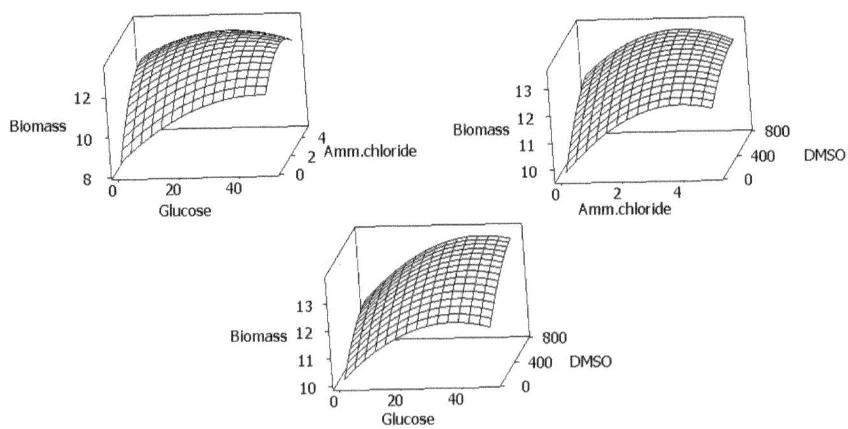

Fig. 1 Surface plots showing interactive effects of glucose, ammonium chloride and DMSO on biomass yield of *Arthrobacter sulfureus*

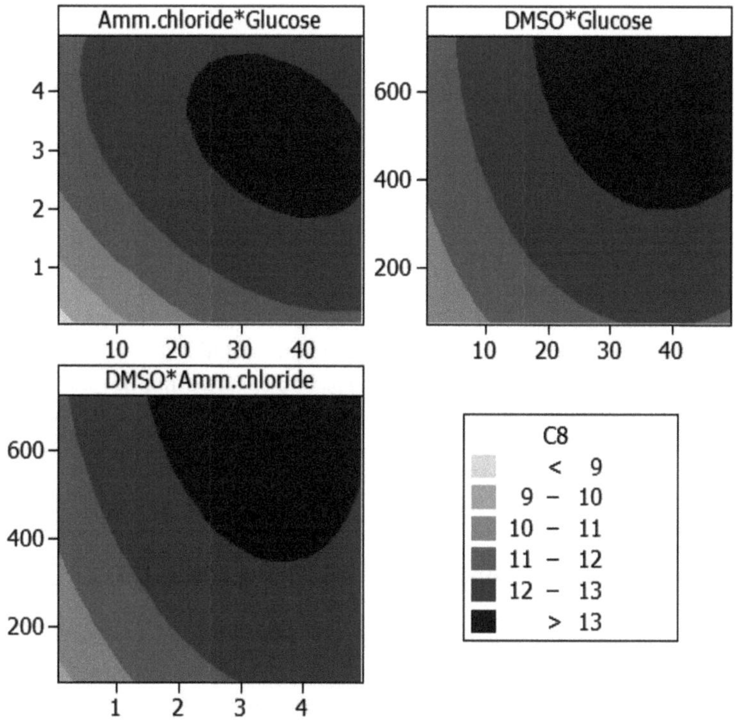

Fig. 2 Contour plots showing interactive effects of glucose, ammonium chloride and DMSO on biomass yield of *Arthrobacter sulfureus*

experiment performed with the concentration of variables after optimization gave a biomass yield of 13.49 g/L and hence the effectiveness of the design at optimized conditions was verified (Fig. 2).

3.3 Biodesulphurisation of Model Oil

3.3.1 Effect of Cell Density on BDS of Model Oil

To study the impact of cell concentration of *Arhtrobacter sulfureus* on desulphurisation of model oil, resting cells were prepared which acts as an aqueous phase. Cell suspension of 0.2, 1, 2, 3 and 4 g (DCW) L^{-1} in aqueous phase is mixed with model oil (having 20 ppm sulphur content) at a ratio of 3:1 (water-oil ratio) in 100 ml Erlenmeyer flasks with total working volume of 20 ml for biphasic system. Rotary shaker was used to agitate the flask at 150 rpm, 30 °C and stopped after 5 days. Sulphur reduction was analysed using X-ray fluorescence spectrometer (XRF). As shown in Fig. 3, the sulphur reduction increased with increasing

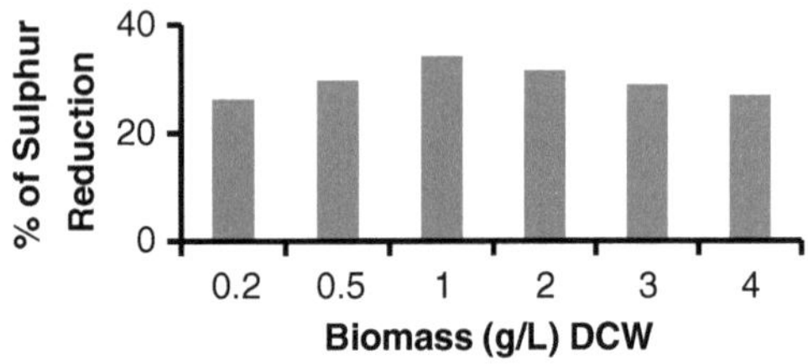

Fig. 3 Final sulphur concentration in two-phase reaction of model oil for different cell concentration

biomass concentration until 1 g/L and later started decreasing at higher biomass concentration, possibility being due to the mass transfer limitation, especially oxygen transfer needed for the oxidation reaction of DBT as said by Maghsoudi et al. (2000) by examining *Rhodococcus* sp. P32C1. This implies that the higher biocatalyst concentration did not mean more DBT molecules converted, and probably the process would achieve better conversion level by increasing interface surface.

3.3.2 Effect of Sulphur Concentration on BDS of Model Oil

Two-phase reaction was carried out for different DBT concentrations of 10–40 ppm having dissolved in the organic solvent (hexadecane), setting the optimised aqueous phase cell concentration (1 g DCW/L) keeping the ratio of water-oil as 3:1. Desulphurisation percentages of model oil were analysed after 5 days of reaction time i.e., incubation on a rotary shaker at 30 °C and 150 rpm using XRF. As shown in Fig. 4. In biphasic media, *Arthrobacter sulfureus* showed lower desulphurisation percentages with increasing concentration of DBT. The reason for lower desulphurisation percentage is due to lower diffusivity rate of DBT into aqueous phase (Caro et al. 2007). In addition, end-product inhibition effects could also be responsible since the potent 4S pathway inhibitor is HBP (Monticello 2000).

3.3.3 Effect of Phase Ratio on BDS of Model Oil

W/O ratio of 0.75, 1.0, 1.5, 3 and 5 are considered at optimum DBT and cell concentration. Sulphur reduction of model oil after 5 days of reaction time is analysed using XRF (volume of the oil is changed while keeping the volume and

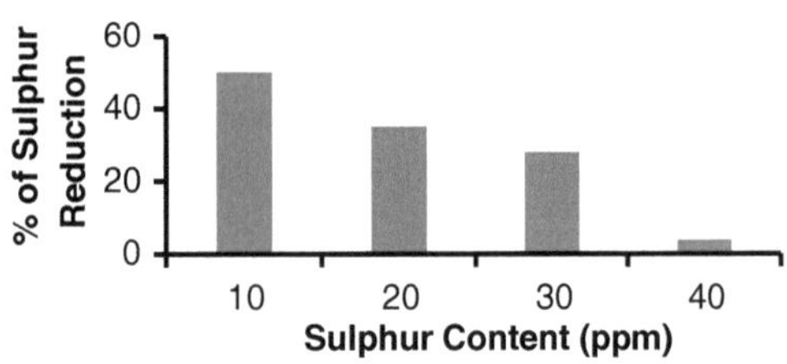

Fig. 4 Final sulphur concentration in two-phase reaction of model oil for different DBT concentration

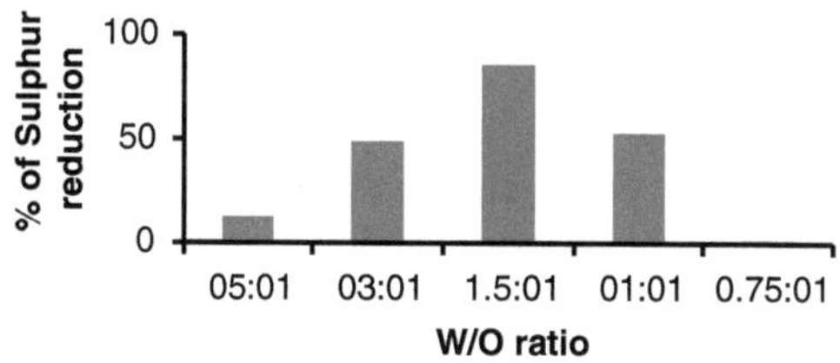

Fig. 5 Final sulphur concentration in two-phase reaction of model oil for different biphasic W/O ratio

the cell concentration of aqueous phase invariable). From Fig. 5, it is observed from the graph that the ratio of 1.5:0.1 showed higher desulphurisation reaction with reduction of up to 82.9 % compared to other ratios. This may be due to mass transfer limitations. Mixing is a necessity for a good mass transfer for creating a satisfactory oil-cell-water contact (Montecello 2000).

3.3.4 Biodesulphurisation of Diesel Oil

Once the optimum values for cell concentration, DBT concentration and phase ratio was obtained for sulphur removal in model oil, sulphur removal from diesel oil was analysed using same optimal conditions as for the model oil. Shake flask studies are carried out by replacing hydrocarbon phase of model oil with high-speed diesel sample (7 ppm). The phosphate buffer (with 2 % glucose) desulphurisation rates as shown in the results were 78.57 % after 5 days of BDS reaction. The decrease in sulphur content of high-speed diesel of 7 ppm (control) to 1.5 ppm shows that the strain has high capability for desulphurisation of diesel oil treated by HDS.

4 Conclusion

DBT has been used as the sulphur compound for growth by most researchers. However, mass production of organism using DBT has been considered commercially impractical because of its high price, low water solubility and growth inhibition by 2-HBP. Therefore, some researchers had previously made an effort to find a alternative sulphur source to DBT such as DMSO. Experimental observations have shown that resting cells of *Arthrobacter sulfureus* grown on DMSO as the sulphur source has higher growth rate than compared to DBT and were able to desulfurize DBT from oil.

BDS process can be efficiently developed by mass production of organism with high desulphurisation activity to be economically feasible. Therefore one can implement an experimental or statistical design like RSM with the objective of identifying the combination of nutrient concentrations that will lead to the highest cell mass concentration, since even small changes in medium formulation is said to have a significant impact on the outcome. The biomass yield of *Arthrobacter sulfureus* obtained after RSM is found to be 13.49 g/L which was previously 7.6 g/L before RSM studies.

Biphasic BDS system is studied for conditions like cell density effects, DBT (substrate) concentration effects and W/O ratio effects, the optimised conditions for the same has been studied and implemented in order to obtain high BDS reaction for both the model oil and diesel oil. Maximum percentage of sulphur reduction for model oil and diesel sample was found to be 85.7 and 78.57 % respectively. Therefore this shows that *Arthrobacter sulfureus* can be efficiently used for BDS reaction.

References

Arulkumar, M., Sathishkumar, P., Palvannan, T.: Optimization of orange G dye adsorption by activated carbon of *Thespesia populnea* pods using response methodolog. J. Hazard. Mater. **186**, 827–834 (2001)

Caro, A., Boltes, K. Leton, P., Garcıa-Calvo, E.: Dienzothiophene biodesulfurization in resting cell conditions by aerobic bacteria. Biochem. Eng. J. **35**, 191–197 (2007)

Gallagher, J.R., Olson, E.S., Stanley, D.C.: Microbial desulfurization of dibenzothiophene: a sulfur-specific pathway. FEMS Microbiol. Lett. **107**, 31–36 (1993)

Hasan, S.H., Ranjan, D., Talat, M.: Water hyacinth biomass (WHB) for the biosorption of hexavalent chromium: optimization of process parameters. Bioresource **5**(2), 563–575 (2010)

Labana, S., Pandey, G., Jain, R.K.: Desulphurization of dibenzothiophene and diesel oils by bacteria. Lett. Appl. Microbiol. **40**, 159–163 (2005)

Luo, M.F., Xing, J.M., Gou, Z.X., Li, S., Liu, H., Chen, J.Y.: Desulfurization of dibenzothiophene by lyophilized cells of *Pseudomonas delafieldii* R-8 in the presence of dodecane. Biochem. Eng. J. **13**, 1–6 (2003)

Maghsoudi, S., Kheirolomoom, A., Vossoughi M., Tanaka E., Katoh S.: Selective desulfurization of dibenzothiophene by newly isolated *Corynebacterium* sp. strain P32C1. Biochem. Eng. J. **5**, 11–6 (2000)

Matsui, T., Hirasawa, K., Konishi, J., Tanaka, Y., Maruhashi, K., Kurane, K.: Microbial desulfurisation of alklated di enzothiophene and alkylated benzothiophene by recombinant *Rhodococcus* sp. Strain T09. Appl. Microbiol. Biotechnol. **56**,196–200 (2001)

Mohbeali, G., Ball, A.S., Katash, A., Rasekh, B.: Dimethyl sulfoxide (DMSO) as the sulfur source for the production of desulfurizing resting cells of *Gordonia alkanivorans* RIPI90A. Microbiology **154**, 878–885 (2008)

Moheali, G., Ball, A., Katash, A., Rasekh, B.: Stailization of water/gas oil emulsions by desulfurizing cells of *Gordonia alkanivorans* RIPI90A. Microbiology (UK) **153**, 1573–1581 (2007)

Monticello, D.J.: Biodesulfurization and the upgrading of petroleum distillates. Curr. Opin. Biotechnol. **11**, 540–546 (2000)

Olson, E.S., Stanle, D.C., Gallaguer, J.R.: Characterization of intermediates in the microbial desulfurization of di enzothiophene. Energy Fuels **7**, 159–64 (1993)

Omori, T., Monna, L., Saiki, Y., Kodama, T.: Desulfurization of dibenzothiophene by *Corynebacterium* sp. strain SY1. Appl. Environ. Microbiol. **58**, 911–915 (1992)

Rhee, S.K., Chang, J.H., Chang, H.N.: Desulfurization of dibenzothiophene and diesel oils by a newly isolated Gordonia strain CYKS1. Appl. Environ. Microbiol. **64**, 2327–2331 (1998)

Shan, G., Xing, J., Zhang, H., Liu, H.: Deep desulfurization of hydrodesulfurized diesel oil by *Pseudomonas delafieldii* R-8. J. Chem. Technol. Biotechnol. **80**, 420–424 (2005)

Tanyildizi M.Ş.: Modeling of adsorption isotherms and kinetics of reactive dye from aqueous solution by peanut hull. Chem. Eng. J. **168**, 1234–1240 (2011)

Wu, J., Wang, J.L., Li, M.H., Lin, J.P., Wei, D.Z.: Optimization of immoilization for selective oxidation of benzyl alcohol by *Gluconobacter oxydans* using response surface methodolog. Biores. Technol. **101**, 8936–8941 (2010)

Yan, H., Kishimoto, M., Omasa, T., Katakura, Y., Suga, K., Okumura, K.: Increase in desulfurization activity of *Rhodococcus erythropolis* KA2-5-1 using ethanol feeding. J. Biosci. Bioeng. **89**(4), 361–366 (2000)

Yu, B., Xu, P., Shi, Q., Ma, C.: Deep desulphurization of diesel oil and crude oils by a newly isolated *Rhodococcus erthropolis* strain. Appl. Environ. Microbiol. **72**, 54–58 (2006)

Effect of Electrodeposited Copper Thin Film on the Morphology and Cell Death of *E. Coli*; an Electron Microscopic Study

Arun Augustin, Harsha Thaira, K. Udaya Bhat and K. Rajendra Udupa

1 Introduction

According to study results from the Centres for Disease Control and Prevention (CDC), hospital acquired infection affects two million people per year, one lacks are dying among them (Reed and Kemmerly 2009; Hassan et al. 2014). Hence, the role of antimicrobial touch surface in hospital has more significance. Even though copper was used against infectious diseases since BC 400, the recent evolution of new microbes renaissance the research of antimicrobial activity and mechanisms of killing microbes by copper (Borkow and Gabbay 2005). The rapid biocidal action of copper was proved against *E. coli* and *salmonella* (Warnes et al. 2012). Many mechanisms has reported for the mode of attacking of copper to microbes which includes plasma membrane attacking by Cu^{2+} ions, copper initiate protein damage and DNA distortion by copper ions (Borkow and Gabbay 2005). But Santo et al. (2011) stated that copper is not targeted the DNA in *E. coli*. Recent studies revealed that copper attacks on the Fe–S clusters in hydratases and thereby inactivate the proteins (Santo et al. 2011; Ben-Sasson et al. 2013; Macomber and Imlay 2009). The understanding of the mechanisms of killing the microbes by the copper is not complete till now. In most of the proposed mechanisms, cell wall is the primary target of attack. In the present study to understand the effect of copper on structure and morphology of gram negative bacteria, *E. coli* has treated with 99.99 % pure

A. Augustin (✉) · K. Udaya Bhat · K.R. Udupa
Department of Metallurgical and Materials Engineering, National Institute of Technology Karnataka, Surathkal, Mangalore, India
e-mail: arunmalabar@yahoo.com

K. Udaya Bhat
e-mail: udayabhatk@gmail.com

H. Thaira
Chemical Engineering Department, National Institute of Technology Karnataka, Surathkal, Mangalore, India

Prasanna B.D. et al. (eds.), *Biotechnology and Biochemical Engineering*,
DOI 10.1007/978-981-10-1920-3_25

copper thin film which is prepared by electrodeposition. The structural and morphological changes of *E. coli* has been analysed by scanning electron microscope and transmission electron microscope.

2 Materials and Methods

5 μm thick copper thin film coating (99.99 % pure) has been done on the commercial pure aluminium of 1 cm^2 surface area by electrodeposition. The freshly coated coupons were kept in the 20 mL *E. coli* ATCC 25922 culture with approximate concentration of 10^7 CFU/mL for 6 h. Equal amount of the same culture was kept without copper as control. The culture was washed in distilled water, centrifuge at 1000 rpm and dried at 35 °C using hot air oven before electron microscopic analysis. The same culture sample has been serially diluted and plated on MHA plate to estimate the number of viable cells from the CFU count. The morphological study of the copper treated and non-treated *E. coli* cell was done by Scanning Electron Microscopy (SEM, JSM-6380, JEOL make). The presence of copper on treated sample was confirmed from Energy Dispersive Spectroscope (EDS) attached with SEM. The structural changes of *E. coli* due to the copper treatment has analysed by Transmission Electron Microscope (TEM, JEM-2100, JEOL make) and the elemental composition of copper treated cell and non-treated cell has been estimated using EDS attached with TEM.

3 Results

Figure 1a shows the SEM micrographs of copper treated *E. coli* bacteria and Fig. 1c represent the EDS corresponding to the same. Compared to the SEM micrograph of non-treated *E. coli* (as shown in Fig. 1b), the treated cells were damaged and wrinkled. Initially the cells were rod shaped as shown in Fig. 1b.

The EDS corresponding to Fig. 1b was shown in Fig. 1d. Compared to normal *E. coli*, EDS results showed that copper treated cells contain 0.1 at.% of copper in the cell. It indicates that noticeable amount of copper ions were present either inside the *E. coli* cell or at the surface of the cell.

Figure 2a shows the bright field TEM micrograph of the copper treated *E. coli* cell. The noticeable change in the cell structure can be observed as compared to the TEM image of non-treated one, as shown in Fig. 2b. The presence of copper in the treatable cell was confirmed from EDS shown in Fig. 2c and that for non-treated cell was shown in Fig. 2d. The TEM micrograph showed that the treated cell has been shrinked to 25 % in size after the copper treatment. The observed distorted and shrunk, rod shape of the *E. coli* is attributed to the leaking out of cell fluid by copper treatment. The number of viable cells calculated from the CFU count shows

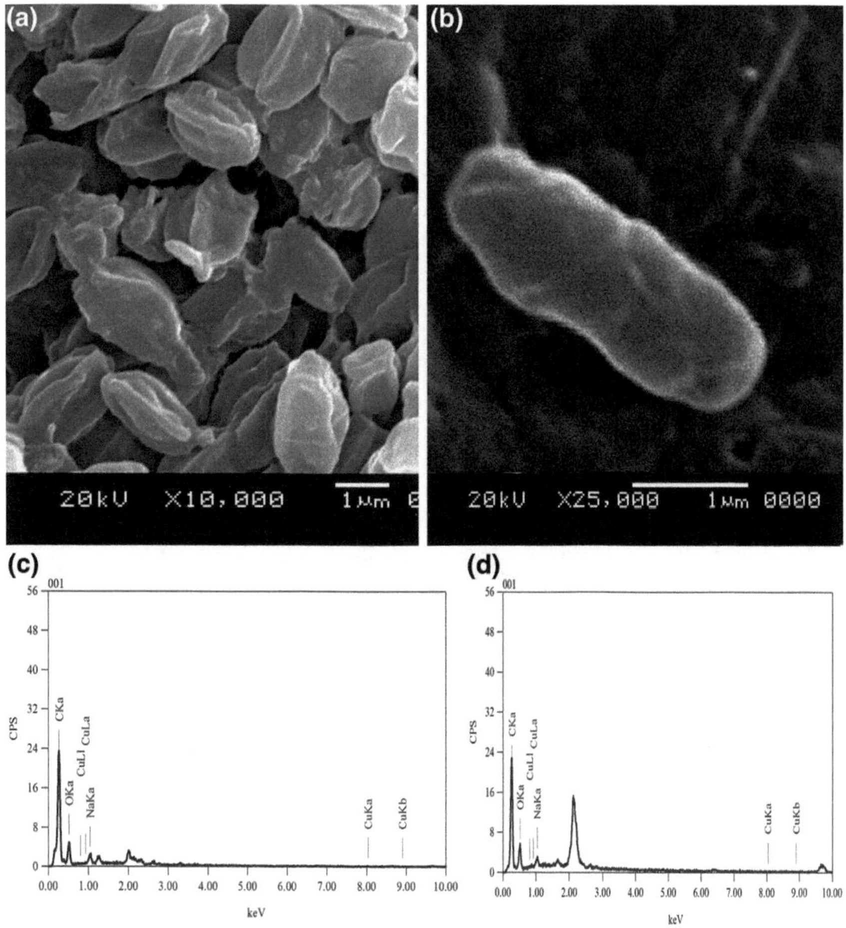

Fig. 1 **a** and **b** SEM micrograph of copper treated and non-treated *E. coli* cell, **c** and **d** EDS corresponding to copper treated and non-treated *E. coli* cell respectively

that 96 % of the cells (viable cell count in control = 22.4×10^6 cells/mL and that of copper coated sample = 8.9×10^5 cells/mL) were dead.

4 Discussion

The electron microscopic images elucidated that the cell membrane as the target of copper attack on *E. coli*. The copper ion treated samples were observed as wrinkled and damaged as compared to the non-treated *E. coli* cells. The wrinkling has happened due to the leakage of body fluids to the extra cellular space. The outer

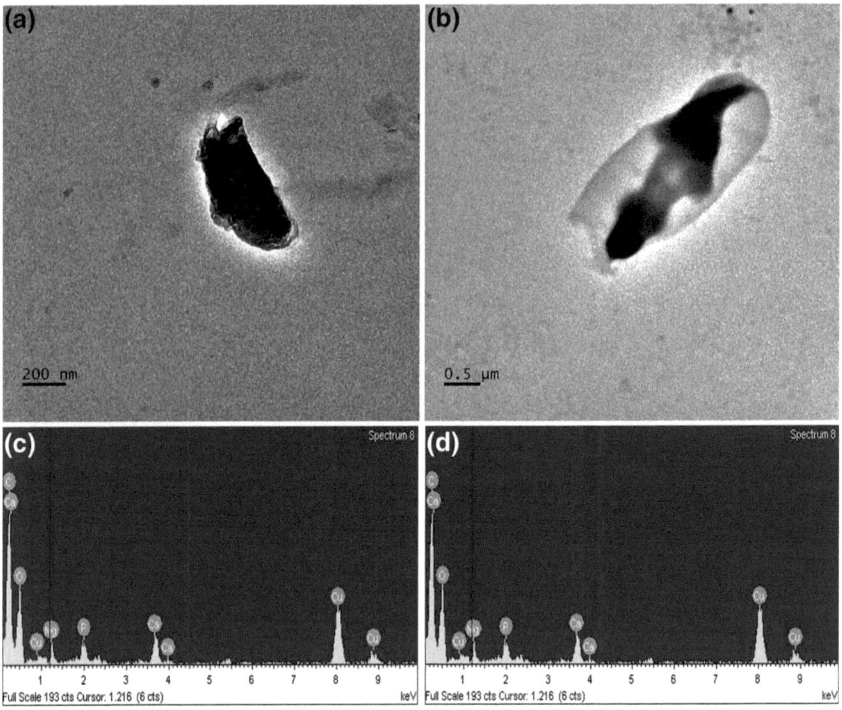

Fig. 2 **a** and **b** TEM micrograph of copper treated and non-treated *E. coli* cell, **c** and **d** EDS corresponding to copper treated and non-treated *E. coli* cell respectively

membrane of *E. coli* is made up of lipopolysaccharides, phospholipids and proteins (Bai et al. 2007). Lipopolysaccharide contains negatively charged groups, which can attract Cu^{+} and Cu^{2+} ions from copper surfaces. These copper ions steal the electrons from the cell wall and affect its integrity. Damage to cell membrane by copper is reported (Santo et al. 2011; Ben-Sasson et al. 2013). When a bacterium comes in contact with a copper surface, a short circuiting of the current in the cell membrane can occurs between positively charged copper ions and negatively charged cell wall. This weakens the membrane and creates holes (Rai and Duran 2011). This disturbs the selective permeability of the plasma membrane. The efflux and influx of the cell imbalanced and loss of K^{+} ions from the cytoplasm reduces its conductivity (Bai et al. 2007). As a result the copper controlling enzymes like CopA, CueO and CuS were inactivated. Moreover the proteins were inactivated due to the breakage of Fe–S bonds in cytoplasmic hydratases like dihydroxy-acid dehydratase in the common branched chain pathway and isopropylmalate isomerase (IPMI) in the leucine-specific branch as reported by Macomber and Imlay (2009). The inherent affinity of copper ions to sulphur makes the rapid cell death and prolonged exposure of cells made the permanent damage of cell wall. Another possibility reported is formation of reactive oxygen species (ROS) at the intracellular region due to the presence of copper in aqueous medium. This generates oxidative stresses in the bacteria followed by DNA damage and finally failure (Shi et al. 2012). It is essential to note that the exact mechanism for microbial damage by copper is not fully understood. Our TEM micrographs indicate a reduction in cell volume to 25 % of initial volume (Fig. 2a, b) and this is possible only if membrane is damaged and cell fluid flows out from the cell. EDS data (Fig. 2c) also indicates the presence of copper ions in the damaged cell. So we conclude that copper is altering morphology and cell death of *E. coli*.

5 Conclusions

Copper induced cell damage of *E. coli* has been observed in the SEM and TEM analysis. The *E. coli* cell which is treated with copper was found to be wrinkled and damaged as compared to the non-treated cell because cell wall was broken, intracellular fluids were leaked out to the extracellular space. The inclusion of copper into the cell was confirmed by EDS attached with SEM and TEM. The Cu^{2+} ions was damaged the cell integrity. Copper ions have broken Fe–S bonds in the hydratases and which leads to the inactivation of cells.

Acknowledgments Authors would like to thank Department of Chemical Engineering NITK Surathkal for providing lab facilities to do *E. coli* culture preparation. Arun Augustin would like to thank MHRD, India for research fellowship.

References

Bai, W., Zhao, K., Asami, K.: Effects of copper on dielectric properties of *E. coli* cells. Colloids Surf. B **58**(2), 105–115 (2007)

Ben-Sasson, M., Zodrow, K.R., Genggeng, Q., Kang, Y., Giannelis, E.P., Elimelech, M.: Surface functionalization of thin-film composite membranes with copper nanoparticles for antimicrobial surface properties. Environ. Sci. Technol. **48**(1), 384–393 (2013)

Borkow, G., Gabbay, J.: Copper as a biocidal tool. Curr. Med. Chem. **12**(18), 2163–2175 (2005)

Hassan, I.A., Parkin, I.P., Nair, S.P., Carmalt, C.J.: Antimicrobial activity of copper and copper (I) oxide thin films deposited via aerosol-assisted CVD. J. Mater. Chem. B **2**(19), 2855–2860 (2014)

Macomber, L., Imlay, J.A.: The iron-sulfur clusters of dehydratases are primary intracellular targets of copper toxicity. Proc. Natl. Acad. Sci. **106**(20), 8344–8349 (2009)

Rai, M., Duran, N.: Metal Nanoparticles in Microbiology. Springer, Berlin (2011)

Reed, D., Kemmerly, S.A.: Infection Control and Prevention: A Review of Hospital-Acquired Infections and the Economic Implications, vol. 9, pp. 27–31 (2009)

Santo, C.E., Lam, E.W., Elowsky, C.G., Quaranta, D., Domaille, D.W., Chang, C.J., Grass, G.: Bacterial killing by dry metallic copper surfaces. Appl. Environ. Microbiol. **77**(3), 794–802 (2011)

Shi, M., Kwon, H.S., Peng, Z., Elder, A., Yang, H.: Effects of surface chemistry on the generation of reactive oxygen species by copper nanoparticles. ACS Nano **6**(3), 2157–2164 (2012)

Warnes, S., Caves, V., Keevil, C.: Mechanism of copper surface toxicity in *Escherichia coli* O157: H7 and *Salmonella* involves immediate membrane depolarization followed by slower rate of DNA destruction which differs from that observed for Gram-positive bacteria. Environ. Microbiol. **14**(7), 1730–1743 (2012)

Optimization of a Glucocorticoid Encapsulated PLGA Nanoparticles for Inflammatory Diseases

Sriprasad Acharya and Bharath Raja Guru

1 Introduction

Age related macular degeneration and retinitis pigmentosa are one of the major causes for blindness all around the world. It is a slow and progressive condition assisted by activated microglia by aggravating internal inflammation. These activated microglia release chemically active molecules that cause cell damage within the outer retina (Lezzi et al. 2012). A viable treatment for this condition is nearly non-existent. Steroids, like prednisolone, have been used as anti-inflammatory agents. Unfortunately these drugs in present dosage lead to severe side effects. Use of biodegradable polymers for encapsulating the drug will not only help in a sustained release but also help in reducing the severe side effects.

PLGA was selected as the carrier since it is one of the most widely used polymer for drug delivery applications (Danhier et al. 2012). The degradation time can be well controlled by varying the copolymer ratio and molecular weight (Vert et al. 1994; Prokop and Davidson 2008).

The drug prednisolone used in the study, is classified as a synthetic glucocorticoid, used in the treatment of several inflammatory and auto-immune conditions. Prolong use is known to cause depression, insomnia, mood swings and memory loss. Thus optimized nanoparticles of PLGA encapsulating the drug prednisolone (Pred-PLGA-NPs) could significantly help in reducing the side effects. Quality by Design is recognized and implemented by regulatory agencies worldwide. The objective of the study was to develop Pred-PLGA-NPs and analyze the effect of the

S. Acharya
Department of Chemical Engineering, Manipal Institute of Technology, Manipal, India
e-mail: sriprasad.acharya@manipal.edu

B.R. Guru (✉)
Department of Biotechnology, Manipal Institute of Technology, Manipal, India
e-mail: Bharath.guru@manipal.edu

Prasanna B.D. et al. (eds.), *Biotechnology and Biochemical Engineering*,
DOI 10.1007/978-981-10-1920-3_26

important factors on the responses like size and drug loading. A 2^3 factorial design was adopted to formulate the Pred-PLGA-NPs.

2 Materials and Methods

2.1 *Materials*

PLGA 50:50 (100,000–120,000) was purchased from Durect Corporation AL, USA. Prednisolone, Polyvinyl alcohol and chloroform were purchased from Sigma-Aldrich.

2.2 *Experiment Design*

A 2^3 factorial design method was employed for the formulations. The variables considered for the study, along with the levels are shown in Table 1. Design expert software (version 9) was employed to design the formulations.

2.3 *Preparation of Pred-PLGA-NPs Using Solvent Evaporation Technique*

Predetermined quantities of prednisolone and PLGA were dissolved in 1.5 ml of chloroform. An oil-in-water emulsion was formed by emulsifying the polymer solution in 15 ml of 1–3 % w/v aqueous PVA solution using a probe sonicator for 2–4 min over an ice bath. The resultant emulsion was then stirred for 18 h at ambient conditions to remove chloroform. Nanoparticles were recovered by centrifugation (15,000 rpm) and washed three times with deionized water to remove excess PVA, followed by lyophilization prior to storage. The formulation composition is summarized in Table 2.

Table 1 The independent and dependent variables with levels

Independent variables	Levels		Dependent variables (responses)
	Low	High	
Drug/polymer ratio (Factor A)	0.2	0.5	Particle size
PVA concentration (%) (Factor B)	1	3	Zeta potential
Sonication time (Factor C)	2	4	Drug loading

Table 2 Formulations and their quantities

Formulation	Factor A: D/P ratio	Factor B: PVA (%)	Factor C: Son. time (min)
P1	0.20	1.00	2.00
P2	0.50	1.00	2.00
P3	0.20	3.00	2.00
P4	0.50	3.00	2.00
P5	0.20	1.00	4.00
P6	0.50	1.00	4.00
P7	0.20	3.00	4.00
P8	0.50	3.00	4.00

2.4 *Measurement of Size and Zeta Potential*

Average particle size (z-average), zeta potential of the developed NPs was determined by laser dynamic light scattering using Malvern Zetasizer (Malvern, Worcestershire, UK).

2.5 *HPLC Conditions*

HPLC was performed using (SHIMADZU Model SPD 20A) equipped with a binary solvent delivery pump. The chromatographic separation was performed on a reverse phase Eclipse plus C18 column, (150 × 4.6 mm, 3.5 μ) and the absorbance was monitored at 254 nm. The system was run with a mobile phase consisting of water +0.1 % TFA: acetonitrile +0.1 % TFA in the ratio of 60:40 (v/v) and was delivered at a flow rate of 0.5 mL/min.

3 Results and Discussion

A set of 8 runs were conducted with 3 independent and 3 dependent variables. All formulations underwent characterization for average particle size, drug loading and zeta potential. The effect of factors on the responses was studied (Table 3) and their respective contour plots were developed.

3.1 *Effect on Zeta Potential*

The zeta potential for nanoparticles helps in assessing the stability of the colloidal systems. The analysis found no significant factors but some unaccounted lurking

Table 3 Response data for the complete set of runs

Std	Factor A: D/P ratio	Factor B: PVA (%)	Factor C: Son. time (min)	Response 1 D (z-average)	Response 2 Zeta pot	Response 3 Drug loading (μg drug/mg of NPs)
1	0.20	1.00	2.00	1696	−3.51	13.042
2	0.50	1.00	2.00	537	−4.64	22.184
3	0.20	3.00	2.00	2354	−13.1	7.184
4	0.50	3.00	2.00	691	−10.4	21.217
5	0.20	1.00	4.00	347	−10.9	5.8
6	0.50	1.00	4.00	503	−14	18.203
7	0.20	3.00	4.00	796	−2.2	9.011
8	0.50	3.00	4.00	539	−11.5	26.892

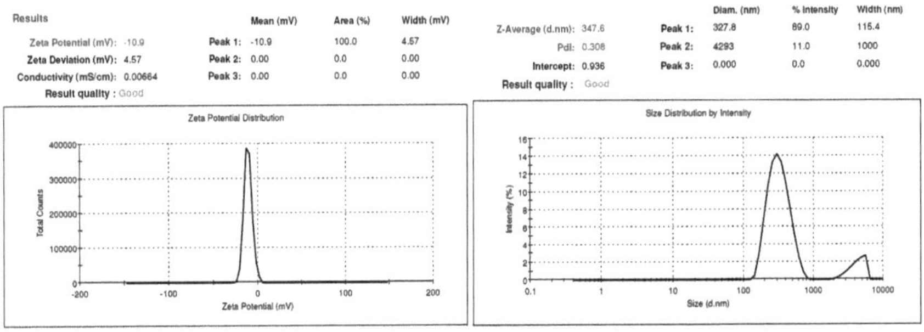

Fig. 1 Zeta potential distribution and size distribution graph of drug loaded PLGA NPs (P5)

variable may have influenced the response during the experiment. The values ranged from −2.2 to −14 mV. Figure 1 shows the zeta potential and size distribution graphs for the drug loaded PLGA NPs (P5).

3.2 *Effect on z-Average*

The factor Drug/polymer ratio (30.35 %) and Sonication time (33.98 %) were found to have significant effect on the size of the NPs. There were no interaction effects. z-Average of developed NPs were in the range of 347 d nm (P5)–2354 d nm (P3) for different variable combinations (Fig. 2).

The model with an F-value of 97.34 was found to be significant ($p < 0.0001$) along with the R-square values, which were in reasonable agreement (Predicted: 0.9345, Adjusted: 0.9857). The observed adequate precision ratio was 27.311, which is well above the minimum prescribed ratio of 4, resulting in a satisfactory signal-to-noise ratio. Thus, this model could be used to navigate the design space.

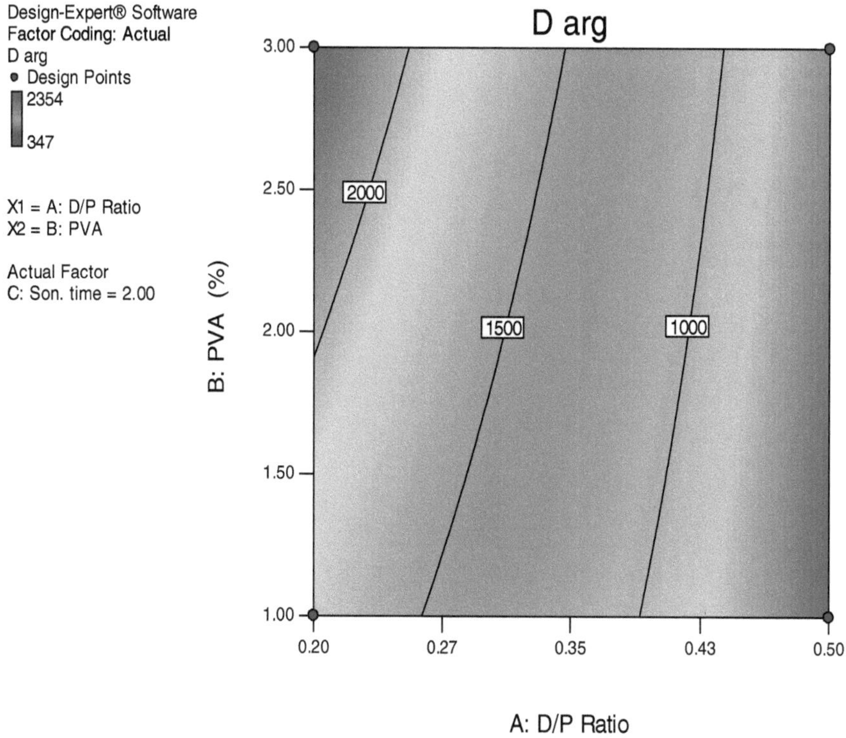

Fig. 2 Contour plot showing effect of D/P ratio and PVA on size of the particle

$$\begin{aligned}\text{z-average} = {} & 4467 - 7710 * A + 429.58333 * B - 1180.25 * C \\ & - 764.16667 * A * B + 2267.5 * A * C \end{aligned} \tag{1}$$

The analysis revealed that the Drug/polymer ratio and Sonication time had a significant but negative effect on the z-average of the polymeric NPs. During emulsification, an increase in drug/polymer ratio may lead to an optimum amount of PLGA being utilized for the encapsulation of the drug (Feczkó et al. 2011; Seju et al. 2011), thus helping in reducing the size of the NPs. The particle size also reduced with the increase in sonication time as finer droplets were being formed by the prolonged and concentrated supply of energy. The factor PVA, showed a slight positive effect as compared to the other factors.

3.3 *Effect on Drug Loading*

The concentration of the drug loaded (μg drug/mg of NP) was found in the range of 5.8 (P5)–26.892 (P8). The model with an F-value of 26.38 was found to be

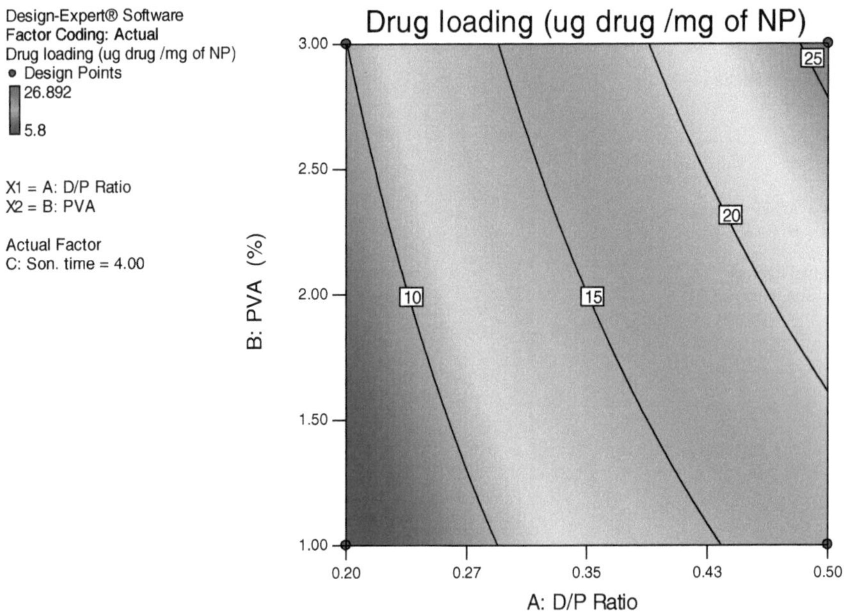

Fig. 3 Contour plot showing effect of D/P ratio and PVA on drug loading

significant ($p < 0.0001$) along with the R-square values, which were in reasonable agreement (Predicted: 0.7610, Adjusted: 0.9477). The observed adequate precision ratio was 12.761, which is well above the minimum prescribed ratio of 4, resulting in a satisfactory signal-to-noise ratio. Thus, this model could be used to navigate the design space.

$$\text{Drug loading} = 20.06838 + 27.26750 * \text{A} - 9.41179 * \text{B} - 5.14638 * \text{C} + 8.64083 * \text{A} * \text{B} + 2.34063 * \text{B} * \text{C} \quad (2)$$

The factor Drug/polymer ratio was found to have the most significant effect (83.89 %) on drug loading. The other factors showed no significant effect on the response (Fig. 3).

The drug loading increased with the increase in the drug to polymer ratio. The drug-polymer interaction and drug miscibility in the organic solution affects the percentage drug entrapment in the NPs (Panyam et al. 2004). An increase in concentration of the drug and polymer in the organic phase causes the viscosity of the organic phase to increase, leading to lesser drug movement into aqueous phase. This may have brought about an increase in the amount of drug inside the NPs (Song et al. 2008).

By applying the constraints (minimize size and maximize the drug loading) on the dependent factors, an optimum Pred-PLGA-NPs formulation was selected. Keeping desirability factor as a basis, Design Expert software was employed to predict the process parameters for the optimized NPs. These Pred-PLGA-NPs obtained by employing the optimized process parameters (0.5:1, 3 % and 4 min) were characterized for z-average (549.8 nm), zeta potential (−15.6 mV) and drug loading (21.169 μg/mg). The response values were in good agreement with the software-predicted values (568.5 nm, −9.950 mV and 25.930 μg/mg), thereby reaffirming the validity of the model.

4 Conclusion

To determine the effects of process variables, a 3-factorial 2-level design and analysis was been carried out. Drug/polymer ratio and sonication time had the most significant effect on the size of the NPs. Maximum levels of Drug/polymer ratio and sonication time showed to produce the smallest particle size. Drug/polymer ratio alone was found to have the most significant and also the largest positive effect on drug loading. The study also found that none of the selected process variables had any significant effect on the zeta potential of the particles.

References

Danhier, F., Ansorena, E., Silva, J.M., Coco, R., Le Breton, A., Préat, V.: PLGA-based nanoparticles: an overview of biomedical applications. J. Controlled Release **161**, 505–522 (2012)

Feczkó, T., Tóth, J., Dósa, G., Gyenis, J.: Influence of process conditions on the mean size of PLGA nanoparticles. Chem. Eng. Process. **50**(8), 846–853 (2011)

Lezzi, R., Guru, B.R., Glybina, I.V., Mishra, M.K., Kennedy, A., Kannan, R.M.: Dendrimer-based targeted intravitreal therapy for sustained attenuation of neuroinflammation in retinal degeneration. Biomaterials **33**(3), 979–988 (2012)

Panyam, J., William, D., Dash, A., Leslie-Pelecky, D., Labhasetwar, V.: Solid-state solubility influences encapsulation and release of hydrophobic drugs from PLGA/PLA nanoparticles. J. Pharm. Sci. **93**(7), 1804–1814 (2004)

Prokop, A., Davidson, J.M.: Nanovehicular intracellular delivery systems. J. Pharm. Sci. **97**, 3518–3590 (2008)

Seju, U., Kumar, A., Sawant, K.K.: Development and evaluation of olanzapine-loaded PLGA nanoparticles for nose-to-brain delivery: in vitro and in vivo studies. Acta Biomater. **7**(12), 4169–4176 (2011)

Song, X., Zhao, Y., Hou, S., Xu, F., Zhao, R., He, J., Cai, Z., Li, Y., Chen, Q.: Dual agents loaded PLGA nanoparticles: systematic study of particle size and drug entrapment efficiency. Eur. J. Pharm. Biopharm. **69**(2), 445–453 (2008)

Vert, M., Mauduit, J., Li, S.: Biodegradation of PLA/GA polymers: increasing complexity. Biomaterials **15**, 1209–1213 (1994)

FSC
www.fsc.org

MIX
Papier aus verantwortungsvollen Quellen
Paper from responsible sources
FSC® C105338

If you have any concerns about our products,
you can contact us on
ProductSafety@springernature.com

In case Publisher is established outside the EU,
the EU authorized representative is:
Springer Nature Customer Service Center GmbH
Europaplatz 3, 69115 Heidelberg, Germany

Printed by Libri Plureos GmbH
in Hamburg, Germany